INTRODUCTION TO ENGINEERING

INTRODUCTION TO ENGINEERING

M. David Burghardt

Hofstra University

HarperCollins*Publishers*

TO LINDA, AMY, AND KATIE

Sponsoring Editor: Don Childress
Project Editor: Janet Tilden
Assistant Art Director: Julie Anderson
Text and Cover Design: Tessing Design Inc.
Cover Coordinator: Julie Anderson
Cover Photo: © COMSTOCK INC.
Photo Research: Karen Koblik, Roberta Knopf
Production Administrator: Kathleen Donnelly
Compositor: The Clarinda Company
Printer and Binder: R.R. Donnelley & Sons Company
Cover Printer: The Lehigh Press, Inc.

Introduction to Engineering

Library of Congress Cataloging-in-Publication Data
Burghardt, M. David.
    Introduction to engineering / M. David Burghardt.
        p.   cm.
    Includes bibliographical references and index.
    ISBN 0-06-041046-9
    1. Engineering—Vocational guidance.   I. Title.
TA157.B86   1992
620'.0023'73—dc20   91-28385
                                      CIP

    93 94 95 9 8 7 6 5 4 3

C O N T E N T S

CHAPTER 4   Ethics and Professional Responsibility   84

CHAPTER 5   Communication—Written and Oral   110

CHAPTER 6   Graphical Communication   130

CHAPTER 10   Mechanical Systems 244

CHAPTER 11   Energy Systems 278

CHAPTER 12   Engineering Economics   314

CHAPTER 13   Computers   340

APPENDICES   377

Solutions to Selected Problems 408

Index 411

Introduction to Engineering is designed to help beginning engineering students decide on their field of engineering, gain a perception of the history of engineering and their place in it, and develop the survival skills necessary for an education and career in engineering. An underlying theme of the text is that society needs people who are educated as engineers to assist in the governance of this technologically complex social system, which in turn requires that engineers become more socially and politically aware and active.

Many students use an introduction to engineering course to better define a field of study about which they are uncertain. They are seeking reasons for studying engineering. A number have preconceived ideas about the profession, though few understand the breadth available to them. Of those students with prior knowledge, some truly want to be design engineers, and most want to earn a good income and eventually to move into management. This latter issue is addressed early on in the text.

Introduction to engineering courses are highly individualized, varying significantly from school to school. Therefore, this text provides the key elements for a variety of introductory courses.

The character of engineering education is changing, reflecting a greater professional concern about the social and political consequences of technological change. Certainly this message must be transmitted to future generations of engineers, and it is one of the philosophical underpinnings of the text, reinforced in virtually every chapter. Not only is the character of the profession changing, but the people entering the profession are different from those who entered it in the past. The stereotypical white male engineer is being joined by people from all races and both genders. The engineering workplace will benefit from the infusion of people with different backgrounds, and our educational programs must address new issues necessary to incorporate these students into the engineering profession.

Often in the past engineering educators presumed students had a personal familiarity with the engineering profession, either through family members or friends. With the new group of students entering engineering education there is often no immediate role model that defines what an engineer is. In this text we attempt to provide such role models.

The underlying purpose of Chapter 1 is to give students techniques for succeeding in college and realizing their potential. Frequently the students who leave engineering do so because of inadequate study habits rather than lack of aptitude or desire. Acquiring good study habits at the outset of their training could help them stay in engineering.

In order to understand what the engineering profession is today, we need to have a historical view of its development. Not only does the material in Chapter 2 trace the evolution of civilized society, but it also describes the evolution of the engineering profession as seen through the formation of the engineering professional societies in the United States in the late 1800s and early 1900s. The dual allegiance to society and to an employer caused much of the conflict in those times and still creates moral dilemmas for engineers today. Furthermore, in the early 1900s engineers were expected to provide solutions to society's problems, and this theme is reemerging today. Perhaps recognizing past pitfalls will help us avoid similar ones in the future.

No introductory course in engineering would be complete without a strong section on ethics and professional responsibility. Chapter 4 is equal to the task and can serve as the springboard for class discussion.

Some engineering majors think of communication (represented by their humanities and social science courses) as not being essential to their success as engineers. Chapter 5 attempts to debunk this view and also supports recent publications of NSPE regarding the need for engineers to be good communicators. Because English composition courses often do not address technical report writing, this topic is covered in Chapter 5 to give students the background they will need to write the many reports required in their undergraduate careers. At first glance it may seem that the topic of résumé preparation is out of place in an introductory course; however, an engineer's career begins with this course, not after completion of the last course during senior year. Library usage is seldom covered in any course, though we often expect students to use the technical resources within the library. The section on library usage has proven to be a valuable asset to students.

An essential element to virtually all introduction to engineering courses is the importance of engineering design. The various aspects of the design process are detailed in Chapter 8. It is here that the creative right side of the brain engages with the analytic left side to yield innovations and technologies that shape our society. In this chapter we look at some techniques and procedures that can help in creating new designs.

Chapters 9 through 13 introduce several generic areas of engineering: electrical, mechanical, and energy systems, engineering economics, and computers. Each chapter has unique features that highlight the topic area and explain why it is relevant to engineering, stimulating the desire to study these fields. Such features include alarm systems, acid rain formation and the greenhouse effect, elementary structural design, cost effectiveness, and computer logic and operation. There are many examples and ample end-of-chapter problems reflecting real-life engineering calculations.

The overall focus of the text is to interest and excite students about the various opportunities afforded by an engineering education: the creative challenge and reward of engineering design, the use of analytic skills in problem formulation and solution in engineering and management, and the skills and attitudes necessary for a rewarding and stimulating college education and for a career after graduation.

INTRODUCTION TO ENGINEERING

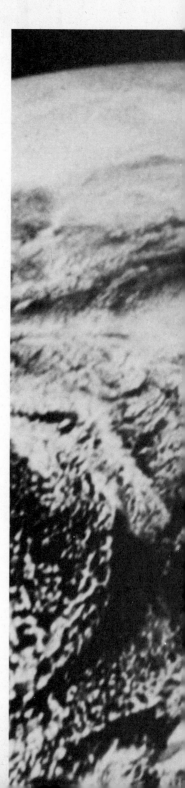

SUCCEEDING IN ENGINEERING

C H A P T E R   O B J E C T I V E S

- To learn what engineering is and is not.

- To identify creativity as an essential element of an engineering education.

- To distinguish between the concepts of technology and engineering.

- To understand why your course of study in engineering is structured as it is and contains the courses it does.

- To make use of techniques that will minimize and improve your study time.

(Photo courtesy of NASA.)

You are on the threshold of a very exciting career in engineering, at a time when the world needs the best technologically aware and creative people possible. People who can solve the multitude of environmental problems that confront us and who can bring us closer together in a world community will lead our future.

1.1 INTRODUCTION

What is engineering? How do you succeed at it? You are embarking on a potential career in engineering, and with the help of this course and text and all the other courses and texts you will have, a definition of engineering will evolve. It may be one of the most useful majors a student can select among all the choices available at most colleges and universities. It is a course of study followed by a professional life devoted to the creative solution of problems. To succeed in any endeavor you must fully utilize your natural abilities, and in the end you will judge your own success on how well you accomplished this. The world that envelops us is a technological one. Think about all the technology you encountered before you left home this morning.

The alarm clock wakes you. Not only is the clock run by electricity, which must be produced somehow, somewhere, but someone, somewhere, designed the alarm mechanism.

You take a shower. The faucets are valves, which were designed by an engineer. The water comes through piping from a central water supply. The entire water distribution system was conceived and designed by engineers.

Opening the refrigerator for milk and juice, you realize that the refrigerator had to have been designed and built according to engineering specifications.

You can expand this list and should do so to gain a perspective of how totally our lives are surrounded by and dependent on technology. No longer do we rise with the sun and depend on manual labor in the fields or forest for our livelihoods; a more sophisticated existence is required. If one does not understand the technology that governs so much of our support systems, a sense of unease or helplessness develops. An underlying advantage of your engineering education is that you will understand the myriad technologies that govern so much of our lives, and thus not be controlled by them.

We all have a limited understanding of what an engineer is. Through education and with the practice of engineering our perception of engineering and what it means to be an engineer will expand. The term *engineer* conjures up many images. Who is your favorite engineer? An aunt or uncle, a parent, cousin, or friend? A famous inventor from history? An unusual engineer whose works are enjoyed by many is Alfred Hitchcock. Although he is known as a movie director, he was trained as a mechanical engineer. We will better understand the close link that engineering has to other creative fields later in this chapter. Contrast the crisp sequencing in a movie like *The 39 Steps* or *North by Northwest* with the more ephemeral movie, *A Man and a Woman,* directed by Claude Lelouch. The influence of engineering is apparent.

One of the difficulties we often face when uncertain about selecting a course of action, such as studying engineering, is that others can dissuade us from that course with thoughtless remarks. It is demoralizing to suffer comments such as ''Why do you want to study that?'' or ''That's too hard,'' or even ''Engineers are nerds.'' As we will learn more fully in Chapter 3, part of the reason for this perception is that engineers often do not remain practicing engineers, and hence people do not directly see the value of an engineering education. Would anyone question your desire to follow in the footsteps of engineers such as Thomas Jefferson, Benjamin Franklin, David Sarnoff, Neil Armstrong, Lee Iacocca, or Jimmy Carter? Society needs people educated as engineers to help solve the problems that daily confront our lives. Historically engineers have been trained not only to solve technical problems but also to have the techniques and the understanding at their disposal to solve nontechnical problems as well. This will be part of your challenge in life. This will provide your response to provocative comments, your reason why the hard work of an engineering education is well worth the effort.

1.2 CREATIVITY IN ENGINEERING

One of the reasons a career in engineering can be so rewarding is that you are compensated very well for creating. Creating is one of the fundamental acts of our existence that gives us great pleasure. Observe small children, or reflect back on your own early childhood, and see the excitement and the development of self-esteem that creating provides. Children are constantly making things—necklaces, castles, cars, rockets—in endless variety. As we mature we become more self-aware and self-critical, fearing to create because others will think it not worthy. The effect of peer pressure in our teens can be very detrimental to creative development. The desire to create remains, however. Your engineering education encourages you to be creative in designing a new product, be it a building, a machine, an electronic circuit, or computer software. With the techniques and knowledge gained in your course of engineering study and with your innate creative ability, something that never existed before can be designed—created. Thus, the general purpose of your engineering education

Art or engineering? In actuality this is a water tower used in a municipality's water supply system. (Courtesy of Sidney B. Bowne and Son)

will be to provide you with technical tools and to encourage the use of your own creativity, culminating in a new design that will solve a given problem.

Engineering is strongly associated with other creative fields, particularly the fine arts. Chapter 2 analyzes engineering from a historical perspective, but here let's leap back in time for just a moment to more primitive eras, before recorded history, when engineering existed in a very fundamental form. We must look to the beginnings of civilization and the forms of creativity exhibited.

How did people create? What is the definition of art? Art is simply something we create, the results of our creativity. Fine art, often confused with the more general term *art,* is a manifestation of creativity with no functional purpose, only aesthetic purpose. Crafts are manifestations of creativity with both functional and aesthetic purpose. The originators of engineering were craft people, such as flint knappers, whose understanding of different types of material, such as flint, and its intrinsic properties allowed them to refine nature and create weapons and tools. In the fine art of painting, people must learn the technical skills of color usage, different application techniques, painting on different surface textures, and the ability to see what exists, before connecting with their innate creative powers and producing what is called a work of art.

Engineers face a similar problem, in that the techniques are the analytic tools with which we must be comfortable, before connecting with our creative powers and producing what is called a new design. Just as the most difficult part of an art student's education is to learn to see what is, rather than their precognition of what is, the most difficult part of an engineering student's education is to learn and feel comfortable with the analytic tools of engineering. When this is accomplished, connecting with your creative being readily occurs.

1.3 TECHNOLOGY AND ITS POLITICAL IMPLICATIONS

The term *technology* is often confused with *engineering,* probably because engineers create it. Technology is the manifestation of engineering creativity; it results from creativity with a purpose, or engineering design. Consider the problem of sending someone to the moon and returning to earth in 1968. The technology to do so did not exist; it had to be created. Created by engineers. Rocket shielding materials that did not exist had to be created to withstand the tremendous temperature generated by reentry into the earth's atmosphere. Computers were created to control the rocket's engines, which also were new. The list is very long. The point is that new technologies are created all the time by engineers solving a given problem. These technologies may be adapted by other engineers to solve a different problem. Thus, the material advances in ceramics used for shielding rockets are now being applied in internal combustion engines.

Engineers are often involved in solving problems that have political aspects. Consider the following situation. The town you live in has been using a landfill as a way to dispose of its garbage. The garbage is dumped and then buried with sand and dirt. The capacity of the landfill area will soon be exceeded, since the garbage can only be piled so high. Also, the chemicals from the garbage are leaching into the groundwater, and hence this method of disposal can no longer be used.

Two alternative methods are available, high-temperature combustion of the garbage, or recycling many of the garbage components and using landfill or

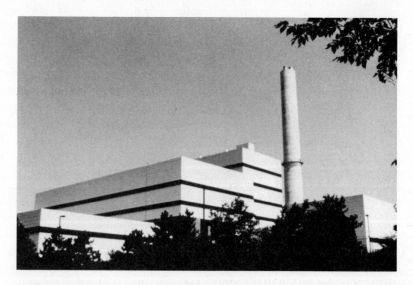

A waste recovery plant incinerates solid waste and generates electricity. This plant has a burning capacity of 2300 tons per day of waste, producing 72 megawatts of electricity. (Courtesy of American REF-FUEL)

incineration of the nonrecycled remainder. The technology for both exists. The heat from the incineration of garbage is used to produce steam, which in turn drives turbines and generators, creating electric power. In this situation the electricity generated partially pays for the incineration process. A major problem is that the incineration of some plastics causes the formation of dioxins, which weaken the immune system of the human body. Metals, such as aluminum, and paper and plastics could be recycled as well. If recycling were combined with incineration, the remaining burnt waste could be buried or used as landfill. The removal of paper and plastic reduces the garbage's heating value, so not as much electricity could be produced, and the cost of the process and your taxes rise. There is also an additional cost in removing the recycled material and selling it, and often little or no market exists for recycled material. Other difficulties are having the populace segregate the garbage so it can be recycled, the cost of developing a recycling machine, and finding a market for the recycled material. In 1985 in New York State, two-thirds of the plastic bottles collected under the bottle deposit system were buried, rather than recycled, because the market conditions made this the most cost-effective alternative.

What exists is a sociological problem needing a technical solution. Research is underway to find a method of combustion of garbage, with plastics included, that will prevent the formation of dioxins, or remove them if they are formed before they are released to the atmosphere. One choice is to go with incineration, and hope that a technical cure will solve the dioxin problem, or accept the limited amount of it as an acceptable risk; a second is to set up a recycling system that requires consumers to separate materials at the pick-up point, or install machinery to do this at a central depot, and have a market for the recycled material. You certainly would not want to stockpile the material. Even in the recycling and incineration system there will be some plastic burned and some dioxin formed, but significantly less than in the straight incineration system.

Both systems require engineering analysis, not high tech engineering necessarily, but traffic flow, storage of materials, plant location, incinerator design. While your mind is focused on this garbage disposal problem, think about the political and technical problems associated with plant location. There is a term that politicians use when everyone wants something, but no one wants it near them—NIMBY, not in my backyard. Engineers must become more attuned to the political aspects of technical solutions.

1.4 WHY DO I HAVE TO STUDY THIS?

All engineering curricula are quite similar in that the latitude available to you in course selection is rather circumscribed compared to nonengineering majors, such as history. Part of this stems from your education being counted as professional experience when it is time for you to be certified as a professional engineer (covered in more detail in Chapter 3) and part stems from the fact that learning the material covered during your senior year requires having taken a sequence of courses in previous years. Engineering requires a vertically integrated education, where more advanced courses depend on information learned in preceding courses. This is contrasted with horizontally integrated curricula, where learning the information in one course does not require the prerequisite of another course. For instance, your understanding the history of modern China does not depend on your understanding Latin America, though, of course, one may enhance the other.

Let's consider the following situation as an example in explaining certain aspects of a mechanical or civil engineering curriculum. A deep stream that is five feet wide needs a bridge so people can walk across it. You find a plank two inches thick, twelve inches wide, and eight feet long and place it on the two banks, crossing the stream (Figure 1.1). No engineering education was needed in this case; from your experience you decided this should·do the job. In fact, you are so pleased that you decide to go into business making plank bridges. But now people who might buy the bridges have questions about how great a load (weight) the bridge can support. Will it support six people? What size people? you ask. And so it goes. From an engineering view, what exists in this case is a beam supported on two ends, and in this simplest view a rigid beam that will not deflect.

You need to analyze this rigid beam. To do this you must model the beam with mathematical equations, which derive from the calculus and differential equations you will be studying. You also need to understand the physical situation, hence the courses in physics you are taking. Then you must take a course in engineering mechanics, where you learn to predict the various forces throughout the beam that are caused by the weight of the people. Finally, for the simplest case, after a course in strength of materials you can determine whether or not a wooden plank with certain assumed properties will break under the aforementioned forces.

Remember at this point you are analyzing what is assumed to be. You are not analyzing the wooden plank that exists: you have developed a mathematical model of that plank that presumes that the wood has certain properties, and from

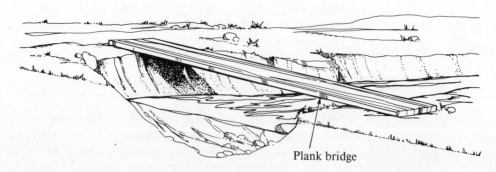

Plank bridge

FIGURE 1.1 A plank bridge across a stream.

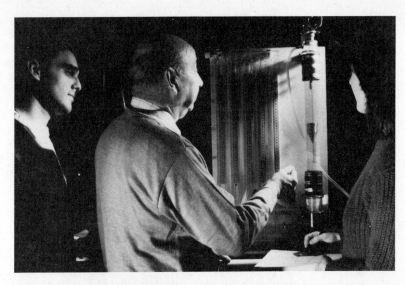

Laboratory experiments often create situations where you can work directly with faculty members, an ideal time for one-on-one learning. (Courtesy of Hofstra University)

this you determine whether or not the plank will sustain the load. You feel certain that it can hold 750 pounds, but to be on the safe side you guarantee it will support 500 pounds. What you have inadvertently introduced is a "factor of safety," which all designs have and which accounts for unexpected load conditions or property variations. We are using the word *design* a little loosely in this case, as you did not really design the plank but assumed it to be sufficient. Another aspect of this situation is that although the model will support 750 pounds, the actual plank might not. Your analysis assumed that the wood had certain properties and could be modeled in a certain mathematical fashion. Perhaps a knot located in the center of the plank causes it to break with a load of 500 pounds. With advanced engineering courses, having as prerequisites additional mathematics and engineering courses, you will be able to account for knotholes. An equally important aspect is that what is designed and analyzed as being satisfactory should be what is built. Often the disasters we read about are caused by a difference between the design and the actual construction. If differences do occur in construction or in materials used, the analysis should be performed again to assure that the de facto new design, what was actually constructed, meets the requirements of the work specifications.

Now a competitor enters the picture, as someone else decided that you are not the only one who can lay planks across a stream. To be competitive and create a better product, you wish to design something new. There could have been cost-saving tactics used first, such as shortening the plank, so there would be less support on each bank, or reducing its thickness, or even reducing the factor of safety, but let's not deal with these at this point. Using your knowledge

The construction of the Hoover Dam on the Colorado River was an immense undertaking in civil engineering. During the years it took to pour the concrete, the river had to be temporarily diverted. The dam creates a recreational lake upstream, Lake Mead, and generates electricity. (Courtesy of ASME)

of strength of materials and engineering mechanics, you create a new bridge design; it is modeled and analyzed to see if it meets the load requirements and the factor of safety you wish. Thus, there is a significant difference between design and analysis; both are important aspects to your engineering education but one without the other is not sufficient in the education of an engineer.

Let's go back to the competitive situation that now exists between your business and others in the same field. Being an ethical person, you selected a very conservative factor of safety. Your competitor chose a smaller factor of safety and feels that it is quite adequate; that means his or her bridge, if of a similar design, will cost less and hence be selected. Municipalities and states have recognized this competitive problem and have established building codes, which define the minimum requirements for a variety of construction situations. Very often the contract specifications will detail the requirements that the design must meet in addition to statutorial ones. This means that everyone is operating from the same vantage point. The problem with this is that the codes and statutory requirements are based on existing designs, and hence design innovation can be restricted by these codes. This is particularly true when new materials allow designs that were not previously possible. Again, technical solutions merge with political realities as special interest groups develop which wish to keep the laws the same so they can continue their existence.

As we continue with this story, let's imagine that your designs are creative and provide you with a product that is better than your competitor's. Business expands, and you hire people to work for you—other engineers and people to fabricate the bridges that you design. Are you a business executive or an engineer? Where is your identity?

Let's change the scenario, using an electrical circuit design to illustrate some other concepts. You are an electrical engineer and work for a small business that manufactures Christmas tree lights. In its thirty years the company has not changed the design from that shown in Figure 1.2. This is known electrically as a series circuit, where the lights are connected in a row, or series. When the string of 20 lights is plugged into the wall, a voltage potential, like a pressure, causes electrical current to flow through the lights and back to the plug. This is the simplest design, but there are some flaws with it. If any of the bulbs are defective, then no current can flow through the circuit, an open circuit occurs, as if the wire were severed, and the entire string of lights will not work. It is also difficult to determine which of twenty bulbs is defective. A second design deficiency is that the brightness of the lights varies with the number of lights on the string. The more lights, the dimmer each glows. This happens in your home when you turn on a hair dryer in the bathroom and the lights dim because there is more resistance, load, on the circuit. You are aware of these limitations and propose a new design, as illustrated in Figure 1.3 (the bulbs are represented by sawtooth lines). In this case the lights are connected in parallel, and whenever a bulb is defective, the others remain lighted. Furthermore, the voltage across each is the same and the brightness is the same, regardless of how many bulbs are added to the string. Your design has a greater reliability than the previous one. A failure of one of the circuit elements does not prevent the other elements from

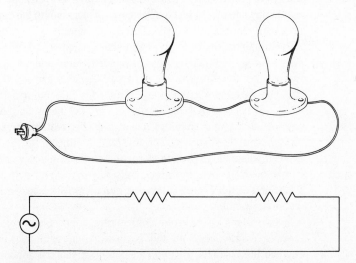

FIGURE 1.2 An electrical circuit with resistors (lights) connected in series.

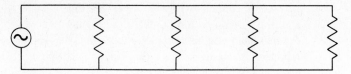

FIGURE 1.3 An electrical circuit with resistors (lights) connected in parallel.

working. Reliability is an important design concept and moves us in the direction of product quality, a topic discussed in Chapter 4. (As your course of study in electrical engineering guides you through very complex circuits, you may want to think about designing a circuit that will cause the lights to blink on and off, randomly or periodically.)

1.5 YOUR CURRICULUM

Why am I studying these courses? Do I really need these courses? These are questions that you undoubtedly have asked or will ask as you pursue your education. Of course they cannot be answered specifically here, but in general the following will give an overview from which you can assess the curriculum of your major.

Your engineering curriculum provides you with the mathematical skills and understanding of the natural sciences that you will need in your later engineering course work. Usually your freshman year is devoted to learning calculus, chemistry, and physics as well as courses introducing the engineering profession, graphical representation, and computer programming. Additionally you start to fulfill requirements in English and social science and humanities; these courses are important for increasing your understanding and awareness of the sociological and political implications of technological change.

The freshman year is very similar for most engineering majors, and it is often not necessary to decide on a specific major until the freshman year is completed. Your curriculum has a flow: the mathematics is necessary for the physics and chemistry, the physics and chemistry for the engineering courses. More mathematics is required for the engineering courses, which is why your course of study is so mathematically intensive. You will model nature with the math, and as your models become more sophisticated, so must your mathematical understanding.

Two elements to your engineering courses are engineering science and design. Your courses in the sophomore and junior years will be primarily engineering science; you will be learning analysis techniques. These courses typically include material science, thermodynamics, circuit theory, and engineering mechanics. In your later courses you will learn more analysis and design

Students often work in teams when performing laboratory experiments. (Courtesy of Hofstra University)

and will make decisions in the problem solution. An engineer must be able to choose among alternative solutions, based on a set of criteria, and your education tries to give you this experience in the later courses, such as machine design, thermal engineering, microprocessor systems, and structural design.

Additionally, you will have technical electives to tailor your course of study to the area of greatest interest to you. Perhaps as an electrical engineer you will specialize in communications, as a mechanical engineer in controls, or as a civil engineer in structures. Most often you will have a capstone design course or courses, in which you will design a product or process from start to finish. In aerospace the aircraft design course considers the optimized design of an aircraft meeting the specifications of payload, range, cruising speed, and runway length. The aircraft characteristics, such as wing shape, are determined and then analyzed and refined. Similar courses exist for other majors, and aspects of design are included in most senior-level engineering courses.

1.6 MINIMIZING YOUR STUDY TIME

Few people like to study; there is always something more exciting to do. For students, however, studying is part of the job, a necessity for being a well-educated person. If you have to study, and sometimes it is fun learning new ideas, then at least study as efficiently as possible, creating the opportunity for other activities in your life.

The study habits you had in high school will in general not be good enough for college. In high school the competition was different; not all students were college bound. Now that you are in college, the average ability of your peers is higher, so your expectations of yourself must improve. One of the differences between college and high school is the lack of outside forces, such as teachers and parents, encouraging your studying. Also, the assignments are less repetitive, so there is less chance to ignore certain material, lounge about, and expect to pick it up later. This twofold difference, no external encouragement and less repetition, is often why freshmen have a difficult time adjusting to college life. For engineering students this is particularly trying, as the material within a course is often vertically integrated, depending on the previous day's assignment, and without persistent attention it is very easy to fall drastically behind. When this happens, you feel discouraged and tend to want to change majors to something easier, though not necessarily better, for you. One of the requirements for surviving an engineering education, and a requirement that businesses are delighted with in all their employees, is that of persistence. There is a trite but true saying, that engineering is more perspiration than inspiration. Your course of study and career will require you to be diligent and persistent; you need not be brilliant to be a fine engineer.

One of the best mechanisms to help you allow enough time to study is creating a schedule for yourself, allotting time for a variety of activities. You cannot expect to eat, go to class, and study as your total life at college. An important aspect of your education is participation in extracurricular activities, which may or may not be associated with engineering. This is particularly important in that engineers must take a more active role in social and political functions. We develop these abilities and interests through contact and practice in college. You might be interested in student government, for instance, and volunteer to work on committees. Perhaps the dormitory you live in needs help in organizing and running activities for its members. Join and participate. However, you need to schedule these activities and your work activities (study) as well. The system you are creating is complex, and you need to manage it well.

You could use an $8\frac{1}{2}'' \times 11''$ piece of paper with the hours of the day and the seven days of the week listed; Figure 1.4 shows an example. It may seem redundant to say the seven days of the week, but many students tend to think of school as a five-day-a-week job. It isn't at all. You must schedule the entire week, as weekends will frequently be spent in report writing and studying.

It is not as bad as you may think. Just think, there are 24 hours to the day, 168 hours per week. Assume that you spend 11 hours a day sleeping and eating, which leaves 91 hours for classes and studying. First, insert your class schedule. Then look at the time around the classes for other required activities — exercising, commuting, team practice if you are involved in sports. You should estimate that, for each class hour, you will be spending or should allot two to three homework hours. Lay this out on the schedule. Suddenly, there are gaps in the schedule where you are not doing anything. Often this occurs between classes; look to use this time effectively. ''Effectively'' may mean attending to student organization work, copying over notes, socializing with other students,

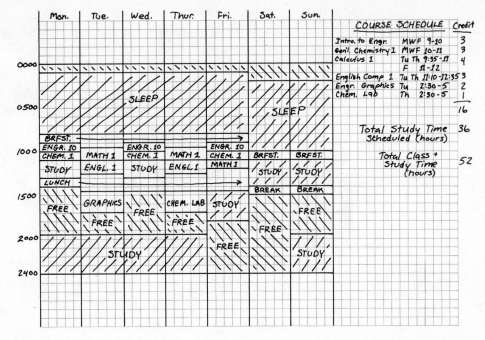

FIGURE 1.4 A chart of the hours in a week, separated into study time, class time, and other activities.

but not focused studying. Whenever you can put together two or more consecutive hours, then you can have a focused study effort. Figure 1.4 indicates this. You do not have to stick to the schedule rigidly, but it should serve as a reminder that if you are going out on Thursday night for a basketball game, the study time has to be made up somewhere in the schedule.

Now that you have a schedule, where is an appropriate place to study? You should choose a place that is well lighted and reasonably quiet and where you will not be bothered by others. It is very difficult to study in your room if your roommate and others are having a discussion. You will be drawn into it and be distracted by it. This does not count as study time. Often the library will have small study areas that are quiet and where the distractions are minimal. Many students believe they must have a television, radio, or stereo playing for them to study. Usually, particularly with television, these create a diversion to your focus. Do not believe you can lie in bed and read something that requires concentration. Sit at a table or desk. Distracting sights and sounds take effort to ignore, and the energy and focus spend in ignoring the surroundings would be better expended in learning the material. This is why the section is labeled "minimizing your study time." Your goal is to learn material as quickly as possible, with as little repetition as possible, to free you for other interesting activities.

Your engineering course work consists of lectures; the responsibility for learning the subject matter is much more yours than it is in most high schools. (Courtesy of Hofstra University)

Part of what you study are notes from lectures. How do you record the lecture information? Through note taking of not only what the instructor writes down, but key words that will trigger your remembrance of those verbal responses not written by the instructor. The notes should not be a verbatim copying of what the faculty member wrote on the board, but rather key words and phrases that when combined with the text material, clarify and expand upon the information in the text. This is one instance where being prepared for class will simplify the task of note taking, as you already know some of the important points and topics. The ability to take useful notes has to be learned. You must listen, comprehend, and take sufficient notes so that with later study you will remember what was discussed and why. It is important to go over your notes after class and amplify them while the material is still fresh in your mind; this is an excellent learning experience. Your notes are a valuable resource if they are legible and you can look at them five weeks later and know what was meant. If you cannot do this, then their purpose is not being met and their value is greatly diminished.

While the fundamental purpose of studying is to gain greater knowledge, a more immediate purpose is preparing for examinations on the material. This is your chance to demonstrate that you have assimilated the information of the course material and translated it into knowledge. The instructor's task is to assess whether and how well this translation has occurred. Cramming, staying up all night before the exam, is not the way to study for a test and is not at all the efficient use of your time. There may be one situation when this type of study can be effective: when you must amass a large number of facts and be able by rote to repeat them on the examination; this does not occur in engineering or technical courses. Rather than deal with this narrow exception, let's analyze how you can best represent yourself to the instructor via an examination. Study the main topics of the course material, which you have done all along in small increments. Study the instructor, see how she presents the material, what aspects she considers important. Your preparation should certainly include these aspects. Pay attention in class; very often the instructor will indicate topics that may later appear on an examination. If you are too busy writing everything down in class, you will probably not be aware of these clues.

Ascertain what type of test it will be. If essay questions are asked, then you will direct your attention to summarizing important concepts, noting trends. You will want to prepare outlines of these concepts so you can form intelligent and thoughtful answers to questions about them. Should the examination be of the short-answer variety, the same information will be asked on the examination, but the relation between key words and specific information is stressed. Most often in engineering courses there are problems to be solved on the examination. Your solving of the homework problems and understanding their solution is the main way to prepare for this type of examination. Your homework must be sufficiently neat and detailed so that when you review it, the difficult steps or intuitive leaps stand out. You should examine these problems and understand the principles you are using. The principles are few, but the variables can be altered in many ways. For instance, in the case where there are three interrelated variables, knowing

Your course work will require the use of computers. From word processing to equation solving, computers are integrated into the fabric of today's engineering curricula. (Courtesy of Hofstra University)

two allows for finding the third. Make sure you understand all the combinations possible and do not rely on the instructor asking the same problem as in your homework, but with different numbers. This is an unlikely occurrence.

Remember that tests are also a learning experience, regardless of the grade. Usually we understand and learn from our mistakes, so the next time they will not occur, thus gaining a more fundamental understanding of the material.

REFERENCES

1. Brown, L., et al. *State of the World 1987*. Norton/Worldwatch Books, New York, 1987.
2. Brown, L., et al. *State of the World 1988*. Norton/Worldwatch Books, New York, 1988.
3. Calder, A. *Calder: An Autobiography with Pictures*. Pantheon Books, 1966.
4. Edwards, B. *Drawing on the Right Side of the Brain*. St. Martins, New York, 1979.
5. Florman, S. C. *The Existential Pleasures of Engineering*. St. Martins, New York, 1981.
6. Metz, L. D., and Klein, R. E. *Man and the Technological Society*. Prentice Hall, Englewood Cliffs, NJ, 1973.
7. *The Technological Dimensions of International Competitiveness*. National Academy of Engineering, Washington, DC, 1988.
8. Waddington, C. H. *Behind Appearances: A Study of the Relations Between Painting and the Natural Sciences in This Century*. MIT Press, Cambridge, 1970.

PROBLEMS

1.1 Write a 200-to-300-word essay on why you are studying engineering.

1.2 Go to the library and, using an unabridged dictionary, copy the definition of the term *engineer*. Amplify this definition with your own ideas.

1.3 Prepare two different weekly schedules for yourself, examining different ways to use your time.

1.4 In your college catalog look at the course of study required for an engineering degree. Note the sequencing of courses in the natural sciences, mathematics, and engineering.

1.5 Make a list of the various student organizations on campus in which you might like to participate. You may need to contact people in student services to find out about these organizations.

1.6 In your home town are there any technical/political issues such as waste disposal, nuclear power generation, or water and air pollution that are of concern? Do you see a role for a citizen-engineer relating to these issues? What is it?

1.7 Visit a local museum, or fine arts gallery, where sculpture is on display. Notice the structure from an engineering view as well as the creativity in the design. Most universities have many pieces of sculpture on display.

1.8 Investigate several engineering projects and determine if they can be completed without using science and mathematics.

1.9   Imagine you were Thomas Edison inventing the electric light bulb. What qualities would you try to create in your invention? Why?

1.10   Describe ten situations that happen in your daily life that force you to be creative.

1.11   Imagine that in your home town a decision has been made to install a monorail public transportation system between two major parts of town. Create the scenario of conflicts, technological as well as political, that might result.

1.12   Design a course of study for the kind of engineer and human being you would like to be.

1.13   Describe the ways in which being an engineer will require knowledge of political science, economics, and psychology.

1.14   Describe your method of studying in high school and consider how you might augment this to deal with the increased expectations of college.

1.15   Imagine the world without electricity. Name ten things that would be more difficult.

1.16   List five ways in which sports have been influenced by technology, for example, graphite-epoxy tennis rackets, metal baseball bats, and fiberglass boats.

1.17   Discuss the questions raised by the changing role of engineers who become business executives or entrepreneurs.

1.18   Discuss the many ways that engineers have affected water, such as dams and water system.

1.19   Describe the attributes of an ideal engineer.

1.20   Look at Alexander Calder's stabiles and mobiles and consider the engineering feat of his works.

1.21   Discuss the impact of electronics on rock music.

1.22   Describe a situation in which you developed a creative solution to a problem. Was your self-image enhanced by the experience?

HISTORICAL
PERSPECTIVE

- To grasp the ramifications of technology on past lives and civilizations.

- To learn the major transformations that have occurred in Western civilization.

- To understand the coalescing of a variety of factors that created the industrial era.

- To trace the formation of engineering professional societies and the development of the engineer's identity.

(W. C. White photo from The Library of Congress.)

Very often we tend to think of engineering in terms of the present day and fail to realize its importance throughout human existence. This chapter will help in changing this perception.

2.1 INTRODUCTION

Why should we examine engineering innovations from centuries ago when there is so much to learn about the present day? We can develop an understanding of the ramifications of technology on our lives today by seeing the effects of earlier technologies on the society of that time. Predicting effects is very difficult and iffy, whereas developing a sensitivity to potential effects is not as difficult and is what we hope engineers possess. There are several ways to study history, and one of the most interesting is to analyze the significant watershed events and determine what caused them. The underlying effects are usually fundamental, as when surplus food supply allows a population to increase, which in turn causes a restructuring of society. Engineering is always pivotally involved in the creation of new technologies, which in this case allows the food supply to increase and leads to the creation of additional technologies permitting societal reorganization.

Another aspect of historical change that we must remember, though it is difficult at times, is the time involved. We tend to look for change in times measured in days, perhaps years at the longest, whereas time in the historical sense is measured in decades and centuries. For this reason our deducing what effects current technological advances will produce is uncertain, but we can be certain that the advances of today will produce changes in how we live and how society is structured in the future.

2.2 PATTERNS OF CHANGE

There have been three major revolutions in the way society has been organized from the beginnings of humankind to the present day. For tens of thousands of years humans lived as hunter-gatherers; the first change occurred when they organized into agricultural communities, and agrarian societies formed the basis of civilization. The second change occurred with the evolution of these societies into industrially based ones. The third change is occurring now, as we move into a post–industrial-based society, sometimes called a computer-based society. What has caused these reorganizations, and what important technologies were involved? We need to examine these questions to understand our roles as engineers, as good citizens, in shaping the society of our progeny.

Human beings tend not to change unless there is some force, external or internal, that directs us to do so. Societal organizations follow the same pattern. The hunter-gatherer groups did not want to change, but the force of hunger directed them to do so. In hunter-gatherer societies the people fed themselves by gathering wild fruits, nuts, and vegetables and hunting animals. Scarcity of food, caused by the end of the ice age ten thousand years ago, forced them to consider forming into agricultural communities, for better control of their food supply. This required an entire shift in the way society was organized; a nomadic tribe or group of people deals with problems differently than a village of people. A different governing structure is needed, different skills are required.

The agriculturally based societies flourished from about 3000 B.C. to A.D. 1300. By this time there were scarcities developing in food and energy supplies throughout Europe. The large feudal estates had produced a surplus of food, the population had expanded and started forming small villages and towns, but the population and the requisite food supply could not be sustained. Famines occurred, and starting in 1348 the black plague swept through Europe repeatedly for a hundred years. The effect of the plague was enormous, as millions of people died. In addition wood was scarce, as more forests were cleared to create farmland. Wood provided the fundamental energy source for the population; it

The earth viewed from space. This gives us a sense of unity, globality, as well as the need to preserve the finite resources of our planet. (Courtesy of ASME)

was burned for heating and used in every manner in forming utensils and implements found in daily life.

A new energy source, coal, had to be used, and a new societal organization, the industrial society, made use of the people who now lived in cities. Industries need concentrated population centers to draw upon. A new manner of understanding the world evolved as well, which set the basis for the scientific method of analysis. The Renaissance took place during this transition from the medieval era to the industrial age. It was a time of turmoil, change, and heightened creativity, not only in technology but in all areas of human activity.

What were the changes in thought, a fundamental shift in the philosophy of life, that occurred in this time of transition that allowed people to make the transition from the feudal, agrarian society to the industrial one? People credit Isaac Newton with discovering the laws of mechanics, which gave rise to our increased understanding of the physical world. He deserves credit for this, but the process started before him. The forces for the creation of the industrial age had been gathering strength for centuries. In 1620 Francis Bacon published *Novum Organum,* in which he espoused an objective way of thinking about the world. This is the way much of western society views the world now, but prior to Bacon scientists and philosophers were concerned with the why of nature. Why did this occur, what was the reason? This is often credited as being the Greek way of studying nature. Bacon on the other hand was concerned with the how of nature. Certain phenomena existed, why did not matter; how could they be explained, and with the explanation how could humans control them? This is a more directed approach and elevates our position as humans to controlling nature, rather than coexisting with it.

It is not sufficient to say this is the way we should think: there must be a structure that allows control to occur. René Descartes provided this structure. Following the publication of *Novum Organum,* though not necessarily because of it, Descartes concluded that it is possible to quantify all of nature as matter in motion and that mathematics holds the key to understanding. Mathematics is the structure needed for the "how" philosophy. Much of nature can be so categorized. You, for instance, have a certain size and exist at a certain point on the earth's surface. This can be developed into mathematical models; peoples' emotional side cannot be so readily modeled, however. Cartesian coordinates, named after Descartes, are used extensively in defining where objects are at a given moment. At this point Isaac Newton entered and provided the means to use the mathematical models. With Newton's laws of motion describing movement in a gravitational field, it was possible to model mechanical systems and predict their behavior ahead of time—a tremendous advance. However, philosophy had shifted from coexistence to control, from seeing quality to seeing quantity.

We should not leave this discussion with the thought that only scientists and engineers were affected by a quantifiable world view. The change in attitude was reflected in economics and history as well. Adam Smith in his book *The Wealth of Nations* tried to develop a set of laws to parallel Newton's mechanical ones. His belief was that material self-interest was the driving force for all individual behavior as well as the behavior of the collective individual, the nation-state, and it is through Smith that the concept of laissez-faire was advanced. John Locke developed a similar view of the function of government. Thus, all facets of society undergo a shift in function and organization as a new structure emerges.

Part of the excitement of being alive now is that this too is a time of great change, a new renaissance with tremendous creative potentials at play and with the possibility of a new societal structure forming that we can contribute to. Engineers have always been the creators of the tools that allow the new structures to form.

2.3 BEGINNINGS OF ENGINEERING

Let's examine a little more systematically and specifically the changes in societies that occurred as we moved from being hunter-gatherers, to agrarian and to industrial society.

How can we say when engineering first began, as there were no records and engineering, as we conceive of it today, did not exist? We will assume that whenever there was an invention or innovation, engineering was required. Thus, engineering really begins with the first humans.

Egypt and Mesopotamia

Humankind existed for tens of thousands of years, organized in small, primitive nomadic tribes and hunting animals and gathering fruits and vegetables. Figure 2.1 shows a hunter using a spear thrower, the result of early engineering. The spear thrower effectively lengthens the arm and gives a greater force to the thrown spear. About 10,000 B.C. the last ice age ended, causing the glaciers to retreat and the summer temperatures to rise. The increased temperature meant a decrease in rainfall and vegetation, and the animal population in the high grasslands migrated to river valleys in search of water and food. Humans followed. They followed the animals into the valleys in northern India, Syria, Central America, and Egypt. Undoubtedly there are other places of early settlement as well, but the archaeological remains have not been discovered. In Egypt the nomads found the river valley of the Nile; in Syria the Tigris and Euphrates rivers were the lure. The geography of the rivers, their flooding times, and the fertility of the soil directed the type of society that would be formed.

Lest we jump too far ahead, the nomadic tribes had to develop rudimentary skills in farming to create an agricultural surplus to support the rest of society. The first farming was done by hand, using a pointed stick to till the soil. Eventually animals were domesticated to assist in farming, and a plow was invented. Figure 2.2 shows a scratch plow, which remained unchanged for thousands of years. A yoke had to be developed before the plow could effectively be pulled by oxen. Thus, the scratch plow, a seemingly ordinary technological achievement, allowed Egyptian and similar societies to feed themselves and increase in population.

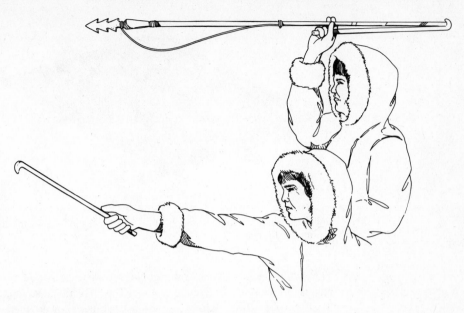

FIGURE 2.1 A spear thrower, used to increase the velocity of the spear by effectively extending the person's arm.

The Nile is formed by two rivers, the White and Blue Niles, the former bringing decayed vegetation to the valley and the latter potash-rich soil. The annual flooding of the river valley occurred in the spring, so the rich soil could be planted and crops harvested in the summer. The population expanded because a surplus of food could be grown in the rich river valley.

There were not enough animal skins to clothe the increased population, so a new material was created: cloth, woven on a loom such as the one in Figure 2.3. The fiber was either animal hair such as wool, or a vegetable fiber such as flax, which is woven into linen.

The agricultural surplus was created because the scratch plow brought greater acreage into production than could have been farmed by humans alone.

FIGURE 2.2 A scratch plow, fundamental to the expansion of an agrarian population. It allowed tillage of much larger acreage for crops than hoeing with a stick.

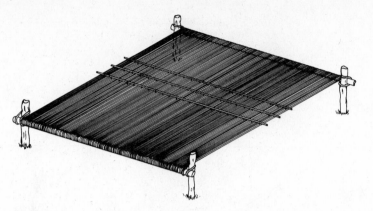

FIGURE 2.3 A loom used for weaving fabric, a replacement for animal skins, for articles of clothing.

The surplus supported the craftspeople, the carpenters, potters, musicians, and bakers, as well as the administrators. The first administrators were probably those people whose knowledge of astronomy allowed them to predict the flooding of the valley, when to plant. Irrigation systems were developed to extend the arable land. Communities formed around certain land divisions, and in Egypt the communities were integrated into a kingdom whereas in Mesopotamia the communities formed into independent city-states. In part because of the annual flooding of the communities, arithmetic and geometry were developed. Land had to be surveyed to locate boundary markers; areas were calculated, weights and measures standardized. Canal construction of the time likewise required development of the same analytic skills.

The tools of construction the Egyptians had were primitive by our standards; the lever, the inclined plane, and the wheel. With these tools and, of course, with thousands of slaves the great pyramids that we associated with ancient Egypt were constructed. The Great Pyramid, with a square base of 756 feet per side, was built within one inch of square. The size is immense; the total number of blocks is about 2.5 million, each weighing two and one-half tons. The construction force has been estimated to have been a hundred thousand men during the flood season, and the construction project took 20 years to complete. The stone was transported from a quarry 500 miles away and probably required the year-round work of several thousand men. The organization, transportation, and feeding of this number of human beings for this long are staggering. It is only with the advent of the industrial age that the use of humans as the primary energy force in construction projects diminished.

This phenomenon was not restricted to the Middle East. The Great Wall of China, which is 3900 miles long including the various branches, though nominally covering a terrain of 1700 miles, was also constructed. The wall is 25 feet high and 20 feet wide, took 400 years to complete, and was started in 3 B.C., about the time of the Roman Empire. Untold thousands of slaves died and were buried in the wall during the construction process.

Thus far humans have had only wood, bone, or stone to work with; how did they make metal objects? Records from about 4000 B.C. indicate that rudimentary metallurgy was understood. Copper ore was reduced to copper metal by placing the ore in a charcoal pit fire. The copper metal melted, settled to the bottom of the pit, and was covered by a glassy slag. The slag was chipped away, and the copper was remelted in a crucible in another fire. The molten copper could be poured into molds. Figure 2.4 illustrates some molds from 3000 B.C.. The molds were crude, and the metal had to be hammered to its final shape. Note the progress from a simple mold in *a* to the more complex molds in *b* and *c*. By trial and error metallurgists of the era found that mixing some tin ore with the copper ore produces a more useful metal, bronze. Bronze has a lower melting point than copper and thus can be more easily worked, but is harder than copper. Bronze was first produced in northern Syria or Turkey, where the tin ore naturally occurred. It was about a thousand years before bronze appeared to any extent in Egypt.

We take the wheel for granted, but it does not occur naturally and had to be created. From about 3500 B.C. there is evidence of the two-wheeled cart, with wheels of wood. Figure 2.5 illustrates how a wheel was cut from a piece of timber. The outer new growth is cut away from the center hardwood. The remaining plank is divided as shown. Indications are that the cart was not used for long-distance transport, but for short hauls, as the axle was attached to the cart by leather straps, so the cart could be dismantled in a rough spot.

The primary scientific advances of the Egyptian and Mesopotamian era were in measurement, surveying, calculation of areas and volumes, and weights and measures. The Egyptians were able to devise a reasonably accurate calendar, probably through their knowledge of astronomy, and elementary use of the right triangle in land measurement. The Mesopotamians developed a method to solve simultaneous equations, though the purpose for learning it is not known. Advances in metallurgy included the development of the bronze alloy of copper and the use of molds in casting weapons, farm implements, and tools. In general the mathematical requirements for running the agrarian societies were not great, and the technologies developed did not require a mathematical skill as a basis for their implementation. Indeed, it was much later in human engineering history that such a skill was needed.

Greece and Rome

The era from 1000 to 300 B.C. represents the development of the Greek culture as we are generally aware of it. Drawings from artifacts indicate that the knowledge of metallurgy was increasing to include ironwork; thus hand tools and weapons could be stronger. The Greeks made advances in sailing and in ship construction and stability; however, the larger ships often were human powered by slave oarsmen. In pottery the Greeks made strides by raising the potter's wheel for greater control and developing an elaborate firing sequence for the kiln. They were able to fire to temperatures of 1000°C. The development of

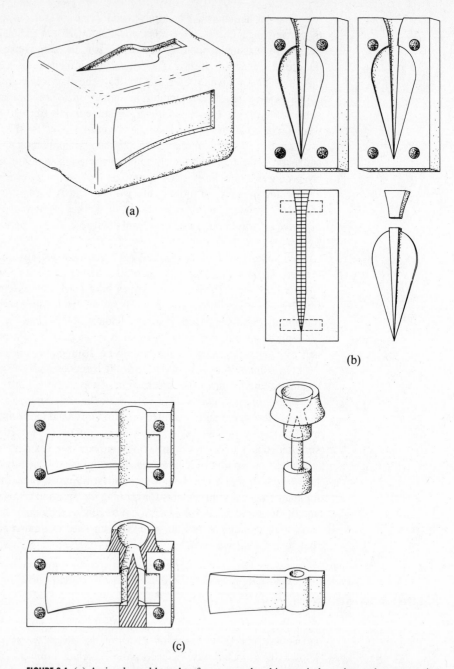

(a)

(b)

(c)

FIGURE 2.4 (a) A simple mold used to form an axehead in one indentation and a spear point in the other. (b) A two-piece mold used to form a spear point. (c) A false-core mold used to form an axehead.

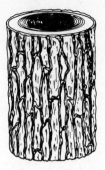 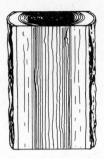 

FIGURE 2.5 How a three-piece wheel is made from a tree. Note that the inner wood, older and firmer than the outer wood, is used for the wheel.

philosophy and of the aesthetic sense were of particular importance to the Greek culture. The Greeks reached conclusions regarding the form of the perfect rectangle; the long side is 1.6 times the short side (the Golden Rectangle). They decided on the column shapes, lengths, and spacings to be used in building construction. This type of restrictive attitude did not lead towards technological innovation, nor was change viewed positively. The Greek view of the world was that it was created in its most perfect form by God. History then is the process of decay and increasing chaos, until finally God restores the world to its original condition and the process starts again. The ideal state was one that allowed few changes, slowing the process of disintegration.

Technological development was not great during the Roman era. Both the Greeks and Romans used the technologies of people they conquered and adapted them to their own use. The Romans were excellent civil engineers; some of their roads still exist. The roadways were constructed by laying a deep subbase of stone followed by a compact base. This method of construction allowed for slow wear, drainage of water, and no heaving of the road in cold weather. During the height of the Roman Empire 180,000 miles of roadway stretched from Turkey to Great Britain. These marvelous roadways were essential to the maintenance of communication between the colonies and the central government in Rome.

The Romans are also recognized for their use of aqueducts in providing fresh water to cities. The aqueducts ran for miles from the country lakes to the urban areas. This required great skill in measurement of distance and elevation to keep the water flowing. The Roman arch, or true arch shown in Figure 2.6, supported the aqueducts and roadways. The figure shows the evolution from an arch requiring a buttress to hold the force exerted by the load on top of the arch. The true arch transmits this force vertically to the ground, eliminating the horizontal component and the need for the buttress, or side support.

A puzzling aspect of the Roman era is the unevenness of the technology. The Romans developed elaborate heating systems for buildings and developed boilers so people could have hot water available on tap, but used a primitive bowl of oil with a wick for lighting. This bowl-lamp was used twelve thousand years before for lighting the caves of the nomads.

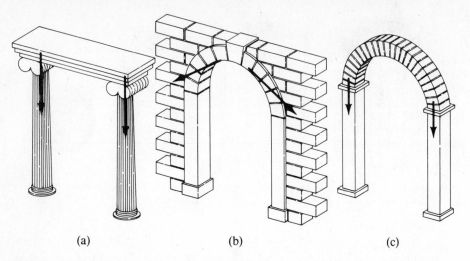

(a)   (b)   (c)

FIGURE 2.6 Diagrams of various arch types: (a) a pillar and lintel arch; (b) a Roman arch; (c) a true arch.

There were societal reasons for the constrained creativity of engineers of this period, concerning what an educated person should have as a career in Greek or Roman times. One could be a philosopher, politician and/or jurist, a general, or all of the above. An engineer could act to enhance the politician's or general's reputation. Thus, building roadways or water supplies and draining swamps increased the politician's standing, and designing better weapons increased the general's status. Mechanisms were developed to do just this. Significant creative ideas did occur, but the follow-through necessary to perfect them was missing. For instance, the Greek Archimedes devised the screw pump, shown in Figure 2.7, and the compound pulley used for lifting heavy objects. He also designed military engines for the defense of Syracuse. The screw pump raised water

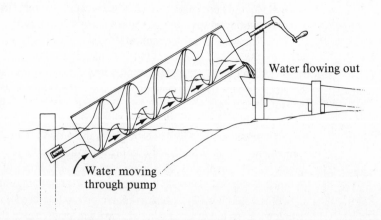

Water flowing out

Water moving
through pump

FIGURE 2.7 A screw pump used to raise water.

continuously and could be used in irrigation, in water supplies, and in draining water from mines. Of course humans had to supply the cranking power—the role of slaves dominates this time period. Releasing people, often slaves, from toil would create unrest, as society had no place for them, so making life easier was not a motivation. Most of the output of inventors was directed towards gimmicks, such as doors of temples opening mysteriously when a fire was lit. The fire would heat air, causing a volume change, and the expansion would, through gears and pulleys, cause the door to open.

The crossbow and catapult in Figure 2.8 were invented by the Greeks and improved upon by the Romans. The Romans also devised the waterwheel; Figure 2.9 illustrates a horizontal waterwheel and Figure 2.10 the more common overshot waterwheel. Despite the inventions the Roman Empire did not last and

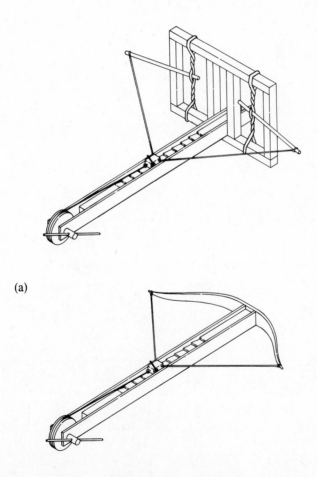

(a)

(b)

FIGURE 2.8 Sketches of (a) a catapult and (b) a crossbow.

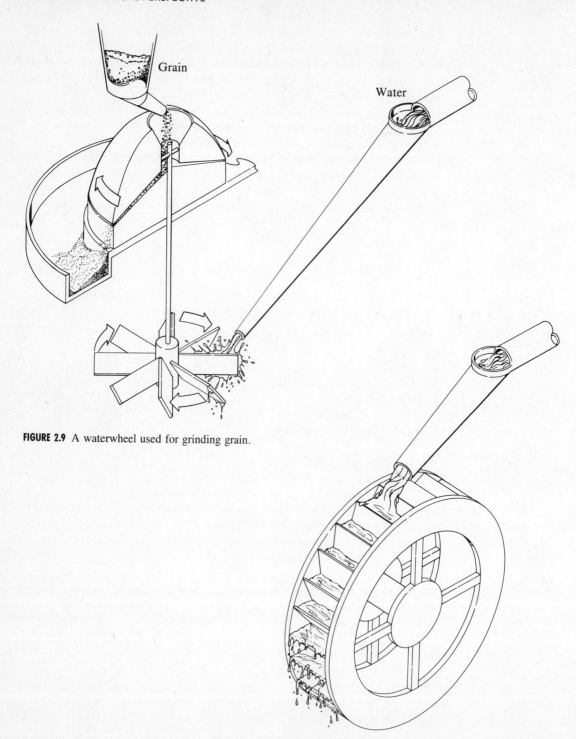

FIGURE 2.9 A waterwheel used for grinding grain.

FIGURE 2.10 An overshot waterwheel.

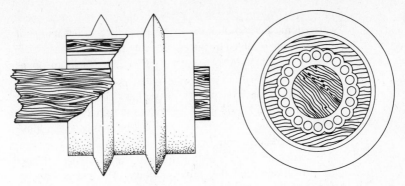

FIGURE 2.11 A wood and bronze roller bearing used to support an axle.

eventually was overrun by what are called the barbarians from the north and east. These people were not necessarily barbarians: after all, which culture was addicted to slavery?

These nomads from the north had independently developed technological advances of some sophistication. For example, Figure 2.11 illustrates a roller bearing that was used about 100 B.C. by people in Denmark. The technology did not last and may have died with the inventor, but it was a significant breakthrough in the support of a rotating shaft. Certainly the seamanship of the Gauls and their vessel design was superior to that of the Romans and was so noted by Caesar. All this is to show that technological inventiveness is not dependent on a civilized culture, but on a creative mind. An accepting culture is necessary for the implementation of new technologies and their improvement.

Dark Ages and Middle Ages

With the collapse of the Roman Empire in the fourth and fifth centuries A.D., what is known as the Dark Ages descended on Europe but not throughout the world. Perhaps the Dark Ages were not really so dark, as it was during this time that animals, and some waterwheels, began to replace humans as the power source. The Arabs were developing paper making, chemistry, and optics, and the Chinese were developing clocks, astronomical instruments, the loom and spinning wheel, and gunpowder. The importance of paper cannot be overestimated, as it allows for the written communication of ideas.

During this period the word *engineer* began to appear. Its root lies in the Latin word *ingeniare,* ''to design or devise.'' About A.D. 200 *ingenium* was used to describe a battering ram, and by A.D. 1200 *ingeniator* was used to describe a person who operated machines of war. Thus, there is a tie-in with inventiveness and warfare in the word and background of engineering. It remained for the industrial age to blossom with the use of mathematics and Newton's laws of motion, for the engineer to combine native inventiveness with scientific theory to create the power of design and analysis available today.

We notice that there has been a continuous association of engineering with domestic and military applications throughout human development. Sometimes we are troubled by the engineering profession's association with the military, and it is something we must each resolve. Seldom do people associate negative qualities with Leonardo da Vinci, the famous artist who painted *The Last Supper* and *Mona Lisa,* but who is perhaps equally well known as a military and civil engineer. He offered his services to princes of Italian city-states as an engineer, designing catapults as well as bridges and buildings. What remains most remarkable about him are his sketches of future engineering devices such as the machine gun, breech-loading cannon, tanks, helicopter, drawbridges, roller bearings, and the universal joint. The list extends to many more ingenious devices. (Notice *ingenious* has the same Latin root as *engineer.*) If you look at Leonardo's career, you can learn more about the engineering profession at the time.

Engineering was divided into two branches, military and civil. This separation persisted until the late 1880s. The funding of engineering works came principally from governments, including feudal kingdoms, which were interested in buildings and bridges and military defensive and offensive weaponry. This has not changed significantly in today's world; a significant portion of engineering design and effort is supported by government funding. The same national self-interest persists in governments today, hence the need for engineers to be employed by governments in much the same way Leonardo da Vinci was.

2.4 THE DEVELOPING INDUSTRIAL AGE

We learned of the forces in society that changed the agrarian structure that had existed since 3000 B.C. to an industrially based structure. This did not happen abruptly or with the contributions only of Bacon, Descartes, and Newton. Rather there were many contributors to new understanding of the physical world, and it is impossible to mention all of them. Assisting all of these contributors was the invention of the printing press and the dissemination of information it allowed. Johannes Gutenberg created the movable-type printing press in about 1454. The alphanumeric characters are cast in lead, and the printer selected the letters to compose words. This task was performed by hand until the early 1900s, when the Linotype machine was developed, a sort of typewriter that mechanically selects the characters and sets them in molds.

In the 1600s Galileo discovered that gravitational acceleration, hence the velocity a body achieves while falling, is independent of weight. In his study of motion, Galileo also concluded that the earth moves around the sun, in contradiction to the view of Aristotle, proposed centuries before. It was this assertion that caused the church to label him a heretic at the time.

The number of people contributing to the advance of science and technology increased each year, as the accomplishments of one fueled intellectual curiosity and competitiveness in others. A student of Galileo, Evangelista Torricelli, linked hydrostatics and dynamics and was responsible, with Blaise Pascal, for the development of the barometer. Robert Boyle discovered the expansion quality of air and the correlation between temperature, volume, and pressure. Robert Hooke discovered that a material lengthens in proportion to the force exerted on it, up to the elastic limit, and in compression it shortens in a similar fashion. Christiaan Huygens developed the spiral watch spring and the pendulum clock and, using his knowledge of the pendulum, was able to measure gravitational acceleration. Isaac Newton and Gottfried Wilhelm Leibniz independently developed differential calculus, essential to the mathematical analysis of most physical systems. In 1698 Thomas Savery patented a steam pump, used in the draining of mines, wherein the condensing steam on one side of a piston would create a vacuum and cause water to be pulled into the other side of the piston. On filling with steam again, the water would be forced out. The valves were opened and closed by hand, and the pump could complete five cycles a minute.

These forerunners and others set the stage for the next century and a half of development. Thomas Newcomen and then James Watt developed the steam engine. Watt's was a superior design, and it helped provide the power needed in textile mills, iron furnaces, rolling mills, and other industries. Other steam-powered machines and devices were created. The steam locomotive was developed once Watt's patent expired, as it required a horizontal steam engine that Watt viewed as impossible.

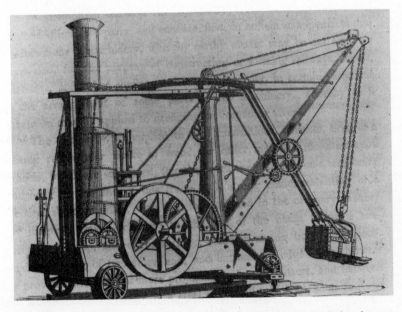

This 1910 steam shovel appears much older to our eyes, though all the elements necessary for its operation remain the same today. (Courtesy of ASME)

While on the topic of engines we must mention the forerunners of the automotive engine. The first was devised and built by Jean-Joseph Étienne Lenoir in 1860 using gas and air. In 1862 Beau de Rochas published an analysis of the processes required of a successful internal combustion engine, and in 1867 Nikolaus August Otto and Eugen Langen developed a gas engine that was more efficient and faster than Lenoir's. In 1876 they designed a four-stroke cycle engine, the precursor of today's automobile engine, using the cycle defined by Beau de Rochas.

The discovery and development of electrical engineering came later in history than its mechanical counterpart. Pieter van Musschenbroek of Leyden developed a device to hold a static electrical charge, now called the Leyden jar, in 1746. This is now called a capacitor, although it was given the name *condenser* by Alessandro Volta. The next step in development came in 1785 when Charles-Augustin de Coulomb developed what is known as Coulomb's law, that between two small electrically charged spheres is a force of attraction or repulsion which is inversely proportional to their distance apart.

Two Italians discovered the principles of electrical conduction. Luigi Galvani in 1786 discovered the reaction a frog's muscle has to electrical stimulation, and in 1782 Alessandro Volta discovered the principals for creating an electric battery. The invention of the battery was a tremendous step, as it demonstrated conclusively that electrical conduction existed. Heretofore only static electricity was thought to exist.

By 1822 André-Marie Ampère had experimentally confirmed the flow of electrical current, leading to the science of electrodynamics. In 1831 Michael Faraday hit upon the means to generate electricity. He found that electrical current was induced in a wire by either increasing or decreasing the intensity of the surrounding magnetic field or by moving the conductor through the magnetic field. It is from these principals that an electric generator is constructed. Concurrently an American, Joseph Henry, discovered the same phenomena as Faraday, but he did not publish his results until 1832. In 1865 James Clerk Maxwell published a paper, "A Dynamic Theory of the Electromagnetic Field," in which he developed equations relating electrical conductivity, electric and magnetic fields, and mechanical force. He concluded, in part, that electromagnetic waves travel at the speed of light. This was confirmed 20 years later by Heinrich Rudolph Hertz. In 1897 Joseph John Thomson discovered the electron, supplying the first physical evidence of its existence in nature.

Before we tire of the list of inventors and discoverers let's mention just one more, Jagadis Chandra Bose, a Bengali engineer and scientist whose initial field of specialization was electrical engineering, but who is best known in the area of plant physiology. In 1895 in India he demonstrated the transmission of electric signals through space, independent of a wire. In 1896 Marconi was awarded a patent for the same achievement. Bose was invited to London in 1897 and lectured before the British Association for the Advancement of Science regarding his measurements of electromagnetic waves. Often the people we remember and who are credited with an invention, or a leap forward, had competitors who were first to the goal but were not recorded. Likewise certain societies have discovered a new technology, such as the manufacture of gunpowder by China or the design of the steam turbine in ancient Greece, but did not use it in the way we expect or it is used today.

Throughout the twentieth century the development of new technology has increased at an exponential rate. The creation of new materials has resulted in a variety of products that could not have been designed and manufactured without them. Nylon, the first synthetic fiber, was announced by du Pont Company in 1938 and revolutionized not only the apparel industry, but all areas of manufacture from automotive tires to mechanical gears. In the early 1900s scientists first learned of nuclear energy; in 1942 engineers and scientists first controlled nuclear fission, which led to the development of nuclear power. Of course, today we are immersed in computer technology: microprocessors control many appliances and monitor the efficient operation and fuel adjustment of automobiles. Yet the first electronic computer was designed as recently as 1945. Also in the 1940s the transistor was invented and the development of solid-state electronics began. The list could go on and on, with jet engines, satellites, and radar, not to mention more mundane developments in building construction and automotive tires.

This realistic statue is not a fossilized professor, but rather a creation by J. Seward Johnson, Jr. This sculpture could not be completed without technological advances in thermoplastic resins, which impregnate the statue's clothing. (Courtesy of Hofstra University)

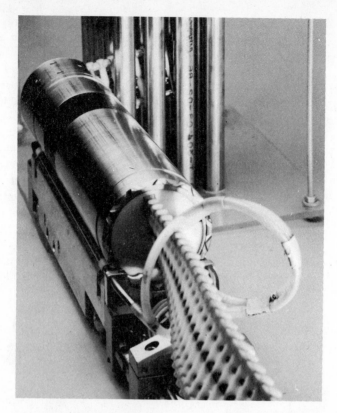

This strange-looking device is a robot, used to inspect and clean the inside surfaces of power plant components, such as steam generator tubes. This is particularly important in nuclear power plants where personnel are restricted from entering these areas unless there is a long shutdown period. A miniature camera attached at the end of inspection ribbon allows the operator to direct the robot's motion and observe the condition of the interior surfaces. (Courtesy of Public Service Electric and Gas Company)

You will be leaders and creators in a new societal organization, dominated by computers and information flow. Computers will affect our lives as never before, especially when the energy crisis of the future, discussed in detail in Chapter 9, arrives. Not only do computers assist us in many of life's daily tasks and work activities—they replace people. In manufacturing, machines can fabricate parts that used to require human assistance. In the area of information flow, computers permit a restructuring of industry; lower-level managers and workers can communicate with those at the highest levels, eliminating layers of middle management. This portends significant changes in career development patterns.

2.5 ENGINEERING SOCIETIES IN THE UNITED STATES

In Chapter 3 we will investigate engineering societies that exist today, but here we will look at the historical roots of these organizations. The development of these societies reflects the conflict or dual allegiance that engineers have towards their profession and their employer.

The roots go back to 1848 with the founding of the Boston Society of Civil Engineers and 1852 with the founding of the American Society of Civil Engineers, in reality the local New York section. Eventually, four other regional civil engineering societies were established, finally affiliating nationally under the auspices of the present American Society of Civil Engineers (ASCE). The purpose of these organizations was to advance the concerns of the engineering profession in all areas, technical and social.

Who were the engineers forming the societies? Many successful engineers were people who developed a business, perhaps a machine shop into a sizable enterprise. We have noted the dilemma facing a person in such a situation: Are you an engineer or a businessperson? Where is your first identity? When the professional societies were formed, the same fundamental conflict existed concerning the functioning of the society: Should it be used to affirm the engineer's loyalty to an employer, that is, the business system, or should it direct the engineer to independent action safeguarding society? The result is that the societies pursue both aims, but this leads to conflict.

The high standards for membership, excluding many potential members, caused a more business-minded engineering society to form in 1871, the

The Robert Hoe Press and Saw Works Iron Foundry, from the 1850s, seems quite primitive compared to today's industry. (Courtesy of ASME)

American Institute of Mining Engineers. As the engineering field became more complex, it was no longer possible for one or two societies to address the needs of the diverse engineering community. Thus, the American Society of Mechanical Engineers was formed in 1880 and the American Institute of Electrical Engineers in 1884. Both of these organizations required meaningful standards to be met for membership and publication, but neither had standards as rigid as the ASCE's at the time, which created a hierarchical arrangement that precluded younger engineers from participation.

We can realize somewhat the changes the new engineers felt as they were the product formed by grafting the newly discovered sciences onto the millennia-old engineering craft tradition. Neither the organizations that were directed towards the crafts nor the science societies were the right fit. The engineering professional societies formed to address this new constituency.

A small group of engineers in the early 1900s had a very positive feeling that the knowledge and abilities that were used in solving problems in the physical world—bridges, machines, and devices—could be adapted and used to solve social problems as well. Herbert Hoover was a leading advocate in this reform movement, and it caused some observers of the time, such as Thorstein Veblen, to portray engineers as the predestined leaders of social change. He wrote many

Advances in technology, created by engineers, remove burden and danger in the workplace. The children pictured here were working in a textile mill in 1910. The affluence created by technology encouraged a more compassionate society that now prohibits child labor. (Courtesy of ASME)

books on economics and society, with his views on engineers reflected in his book *The Engineers and the Price System*. This reform movement united the fragmented engineering societies and perhaps can be credited with presaging scientific management, but it did not succeed in solving social problems.

Thus, engineering and engineers must be evaluated in light of the time they are in. We cannot hope to understand the engineers of the early 1900s if we do not concurrently understand the forces leading to World War I and the politics of the world following the war. Activism was part of the landscape in every nation: the Union of Soviet Socialist Republics experienced its birth, the class system in England was in disarray, the League of Nations was being created. This explains, in part, the need for as broad an education as possible in the social sciences and humanities, so that we may better understand the political, economic, and social environment around us.

Engineers are practical people and understand that businesses need to be financially viable to hire and employ engineers, accountants, machine operators, and office workers and that engineers play a unique and fundamental role in the organization. They create and design new products, which must operate correctly and safely, and at a cost that lets the company remain in business. All this is accomplished in the context of being a responsible professional, being responsible to the engineering profession and to society. It is in this area of societal responsibility, where the engineer has been active in the past, that renewed activity is needed in the future.

REFERENCES

1. Burke, J. *Connections*. Little, Brown, Boston, 1978.
2. Cardwell, D. S. L. *Turning Points in Western Technology*. Science History Publications, New York, 1972.
3. Hodges, H. *Technology in the Ancient World*. A. A. Knopf, New York, 1970.
4. Kirby, R. S.; Withington, S.; Darling, A. B.; and Kilgour, F. G. *Engineering in History*. McGraw-Hill, New York, 1956.
5. Kranzberg, M., and Pursell, C. W., Jr. *Technology in Western Civilization*. Oxford University Press, Oxford, 1967.
6. Layton, E. T., Jr. *The Revolt of the Engineers*. Case Western Reserve University Press, Cleveland, 1971.
7. Mumford, L. *Technics and Civilisation*. Routledge and Kegan Paul, London, 1962.
8. Rifkin, J. *Entropy*. Viking, New York, 1980.

PROBLEMS

2.1 List four examples of technological innovations that have affected the development of Western civilization; list four more affecting the United States.

2.2 Modern technology envelops today's world. Think about the modern supermarket: the door opens automatically, the food is checked out using a scanner. List ten more inventions that we consider everyday parts of life in the supermarket.

2.3 What were the principal reasons that Greek science did not lead to technological advancements?

2.4 Write a brief report on the construction of the Great Wall of China, giving an estimate of material usage and people required.

2.5 Paper is fundamental to our communication of knowledge. Prepare a report on paper manufactured from earliest times.

2.6 Describe Leonardo da Vinci's version of the helicopter and two other of his inventions that you believe are most significant in terms of today's society.

2.7 The first transatlantic cable was laid in 1858. Investigate the technological problems that had to be solved.

2.8 Write a report on the development of the ceramic automotive engine.

2.9 Investigate and write a paper on the technological innovations that may be possible in light of superconductivity.

2.10 Some of the iron railroad bridges that were built in the mid-1800s failed, resulting in loss of life and property. Report on the causes of these early failures.

2.11 The Brooklyn Bridge was designed and constructed under the supervision of John Roebling and his son, Washington. Write a report on the design innovations and the problems they addressed.

2.12 Write a brief report on the discovery of semiconductors and breakthrough investigations of their properties.

2.13 Write a short biography of one of the following, pointing out technical and/or scientific achievements: Archimedes, William Gilbert, Otto von Guericke, Robert Watson, James Watt, Thomas Edison, George Westinghouse, John Gorrie, Robert Goddard, Alessandro Volta.

2.14 Describe how engineering has improved our lives in transportation, food processing, and medicine.

3

FIELDS OF ENGINEERING

■ To learn about the technological team.

■ To investigate the structure and requirements of engineering professional societies.

■ To find out what accreditation of your curriculum means.

■ To distinguish between major fields of engineering study.

■ To become aware of various technical career possibilities within any given field.

(Photo courtesy of NASA.)

In this chapter we will look at some of the more prominent engineering societies, at the types of careers members may pursue, and at the curricula of some typical engineering programs. In addition we will see what it means to be a professional engineer.

3.1 THE TECHNOLOGICAL TEAM

As a professional in the field of engineering you will work with other technically educated people to form what is known as a technological team. The team will bring complementary skills together to solve problems. In the broadest sense it consists of the following groups:

scientists

engineers

technologists

technicians

craftspeople

The education and training for each group of people is different but often overlaps in adjacent areas. Whereas the engineer and scientist both must be well educated in mathematics and science, the scientist's role is to seek out new fundamental understandings of the world, expanding our existing knowledge. A scientist seeks to understand the why of nature. However, very often scientists employed by companies are applied scientists, using the knowledge gained by advances in theoretical understanding to build new products or devices. This is what an engineer does as well.

The technological team usually is directed by an engineer, as the engineering function primarily involves the design of a new product, project, or system. The purpose of an engineer's education is to equip creative minds with the mathematical and analytic skills necessary to conceive of new designs, to intelligently question present ways of accomplishing tasks, and to find better alternative methods, in light of evolving technology.

Whereas the scientist and engineer overlap functions on the abstract end of the spectrum, the engineer and the technologist may overlap at the implementation level. A technologist has graduated from a four-year engineering

Field engineers are taking core samples from a road to determine the subsurface condition. (Courtesy of Sidney B. Bowne and Son)

technology program. These programs typically have fewer mathematical requirements than an engineering program and courses more related to current hardware and processes. The presentation of material is less theoretical than in engineering programs, though the types of courses offered are quite similar. The technologist works with the engineer in all areas, often in the interface between the engineer and the technician, overseeing the technicians' work and further developing the design changes.

Once you are hired by a company, your initiative and abilities determine what assignments will be given you and what your function will be in actuality. The functions of the engineer and the technologist overlap in the workplace, blurring the lines of distinction education delineates.

The technicians are usually graduates of a two-year program in engineering technology. They are "hands-on" people and could be draftsmen, field inspectors on construction projects, or electronic technicians. Craftspeople are skilled workers, historically the first members of the engineering technological team. They have the skills to make devices and systems. Modelmakers are used in the aerospace industry, machinists in a variety of industries, welders and carpenters in the construction industry. Usually they have training in their skill, followed by years of practice.

3.2 ENGINEERING SOCIETIES

In Chapter 2 we saw that by the end of the 1800s the fields of engineering had expanded greatly, resulting in the formation of several engineering societies. This rapid growth in engineering fields has continued to the present day, with the result that there are over 400 engineering societies and related groups in the United States and 21 engineering societies in Canada. This should indicate to you, at least preliminarily, the tremendous diversity in career opportunities that exists.

Several organizations are concerned with the engineering profession as a whole, including the National Society of Professional Engineers (NSPE), the National Academy of Engineering (NAE), and the American Association of Engineering Societies (AAES). The AAES comprises 28 engineering societies and acts to represent the profession, for example in the area of governmental policies and through one of its subgroups, the Engineering Manpower Commission, which undertakes many studies regarding the engineering profes-

These research engineers are developing a sustained fusion reaction at the Tokamak Fusion Test Reactor at the Princeton Plasma Physics Laboratory. (Courtesy of Public Service Electric and Gas)

sion. These studies include research on industrial demand for engineers in the future and the current placement of engineering graduates, and a comprehensive salary survey. The AAES uses this information to support public policies necessary for the continual improvement of the engineering profession.

The beginnings of the unity of purpose hark back to the founding of the United Engineering Trustees, Inc., in 1904 by five engineering societies, called the Founder Societies. They are listed here with their dates of founding.

American Society of Civil Engineers (ASCE 1852)

American Institute of Mining, Metallurgical, and Petroleum Engineers (AIME 1871)

American Society of Mechanical Engineers (ASME 1880)

Institute of Electrical and Electronic Engineers (IEEE 1884)

American Institute of Chemical Engineers (AIChE 1908)

The United Engineering Trustees is the titleholder to the 20-story United Engineering Center which provides office space in New York for 23 engineering societies and related organizations. In addition the Trustees provide to the profession the Engineering Societies Library and related information services and the Engineering Foundation, an endowed foundation dedicated to research in science and engineering directed towards the good of the engineering profession and humanity.

These 23 societies establish goals and disseminate information about themselves. Their objectives are similar to those established by ASCE, listed here.

1. To encourage and publicize discoveries and new techniques throughout the profession.
2. To afford professional associations and develop professional consciousness among civil engineering students.
3. To further research, design, and construction procedures in specialized fields of civil engineering.
4. To give special attention to the professional and economic aspects of the practice of engineering.
5. To enhance the standing of engineers.
6. To maintain and improve standards of engineering education.
7. To bring engineers together for the exchange of information and ideas.

The societies implement these objectives in a variety of ways. At national and regional meetings each society addresses these issues. Often at these meetings professional articles are presented by investigators in the field. A variety of technical journals sponsored by the societies disseminate technical information, and there are student chapters of these societies which you should join now.

There are several reasons for becoming a member while being a student. Contacts with engineers working in industry may be possible. You can participate in running the student chapter, creating an opportunity to speak in public and work as a technical team member. In addition you will receive a society periodical relating what is currently happening in the profession, intriguing you with future opportunities for involvement. An important benefit is that the societies all have loan assistance and scholarship programs, which you become eligible for. Following graduation your continued participation is necessary, so that the engineering profession can continue to advance. We have seen that advances in engineering parallel advances in society; without your professional involvement advances occur more slowly. One method of being involved with technology utilization in society is through your professional society.

In addition to having worthy objectives such as those mentioned, the societies have membership requirements to assure that competent engineers direct the organization. The following is a list of the various grades of membership that ASME has and that are similar to those in the other societies.

STUDENT MEMBER.   Belongs to a student chapter at a school with a curriculum approved by the Accreditation Board for Engineering and Technology (ABET).

ASSOCIATE MEMBER.   Must have graduated from an engineering school of recognized standing, or have eight years of acceptable experience.

MEMBER.   Requires six years of practice if one graduated from an engineering school of recognized standing; otherwise, requires twelve years. Must have spent five years in responsible charge of work.

FELLOW.   Requires nomination by fellow members for an engineer or engineering teacher of acknowledged attainments. Also requires 25 years of practice, and 13 years in member grade. This is an honorary grade.

HONORARY MEMBER.   Requires nomination by membership and election by the board of directors. This grade is given to people of distinctive engineering accomplishment.

EXECUTIVE AFFILIATE.   Not necessarily an engineer, but in a position of policy-making authority relating to engineering; must be one who cooperates closely with engineers.

AFFILIATE.   Capable and interested in rendering service to the field of engineering.

The position you can consider at this point is student member; all societies make the cost of membership minimal, and we have already examined the benefits. One additional benefit to being a student member is that you can automatically become an associate member following graduation. There is no membership fee involved, which there is if you do not choose this path, and you do not have to obtain membership forms and nominees.

In addition to societies affiliated with virtually every field of engineering there are professional societies organized to support women and minorities. These include the Society of Women Engineers (SWE), the National Society of Black Engineers (NSBE), and the National Society of Hispanic Engineers (NSHE). All these organizations provide special scholarships and loans, provide role models, network, and discuss specific problems that these groups may encounter in school and industry.

3.3 ACCREDITATION BOARD FOR ENGINEERING AND TECHNOLOGY (ABET)

One of the requirements for ASME student membership and for most others is that the university engineering program be ABET accredited. What does this mean? ABET's function is the examination and accreditation of engineering and technology curricula. ABET comprises 19 participating organizations and 4 affiliate organizations.

There are 31 program areas that ABET accredits for periods of three or six years. At the end of these periods a fact-finding team from ABET (Engineering Accreditation Commission [EAC] for engineering programs and Technology Accreditation Commission [TAC] for technology programs), made of members of the participating and affiliate professional societies, visits the university and analyzes the course of study for the individual programs. The team examines the quality of course work, the competence of the faculty, and the administrative support and, importantly, interviews students. The team writes a report of its observations and submits it to ABET. Based on this and other factors a decision to accredit and for what time period, or not to accredit at all, is made by ABET, and the school is informed of the accreditation action. This assures you, as a student, that the course of study you are pursuing is relevant, that it meets minimum professional standards, and that when you graduate you will be an asset to the engineering profession and to society. As of 1988 there were 1401 accredited bachelor's degree engineering programs, 292 accredited four-year technology programs, and 468 accredited two-year technology programs. Undoubtedly you are enrolled as a major in one of these programs presently. What follows is an overview of some of the fields of engineering you may pursue upon graduation and the types of projects you can be involved with.

3.4 FIELDS OF ENGINEERING

Aeronautical and Aerospace Engineering

The term *aerospace engineering* is often used to denote the entire field and the aerospace industry, but we should make a distinction between aeronautical and aerospace engineering. Aeronautical engineering deals with flight and the movement of fluids in the earth's atmosphere. Thus, the research, design, and development of all types of airplanes are included in this field. You will specialize in work areas centered on aerodynamics, propulsion, controls, or structures. One of the interesting aspects of aircraft design is the interaction between the air around the plane and the plane's structure: one influences the behavior of the other. If you have been in an airplane, you must have noticed that the wings flex as the airplane flies. They are designed to flex, and their movement affects the air flow around the wing. Furthermore, the development of new materials is important, as materials fatigue after too much flexing and the wings must be replaced. The more durable the material, the greater the wing's life expectancy. Notice the word *material,* not *metal;* much of the materials used in new aircraft design are not metals, but composites of graphite and epoxy.

As we still imagine ourselves aboard the aircraft, someone had to design our aircraft's engines, and has to redesign them as new materials allow different and better operating conditions. For instance, the inlet temperature to the jet's turbine can increase because new turbine blades are able to withstand continuous operation at higher temperatures. Not only is the turbine redesigned for the new conditions, but related equipment is redesigned to operate with a hotter fluid.

While we look to the stars and imagine people rocketing through space, aerospace engineers are required to turn this image into a reality. Quite similar to aeronautical engineering in the job aspects involved, aerospace engineering deals with environments not found on earth—space's vacuum or the atmospheres surrounding other planets. Aerospace engineers must design rockets to leave and reenter earth's atmosphere, structures to exist in space, life-support systems enabling humans to live in space, robotic devices, and control

Perhaps nothing has illustrated a high point in U.S. technological accomplishment as well as the landing of an astronaut on the moon, July 20, 1969. (Courtesy of NASA)

instrumentation. Let's consider just one application for an aerospace engineer, the problem of vibration in an orbiting space station. We seldom think about this on earth, even if the structure is ultralight, as the space station must be, because air assists in the damping of the vibrations. Thus, structures are passively damped, and no vibrations continue for long. (There are exceptions, such as the huge World Trade Center, in New York City, which has a mechanical damping system.) The structures in space must be actively damped; a control system must sense when a vibration is occurring and set up a countermovement in the structure to cancel the vibration, much as a reflected wave will cancel an incident wave.

Agricultural Engineering

For many of us located in urban and suburban communities, the concept of how food is produced is not one we dwell upon, but fortunately for us the agricultural engineer does. A problem since the beginning of humanity is the production of enough food to support the population. Being from an industrialized nation, we tend not to think of this problem, but on a global scale starvation is a reality millions must confront daily. The agricultural engineer blends engineering knowledge with understanding of soil systems, land management, waste management, and environment control to create methods and technologies that will allow the continuation of high crop yields to feed us. There are five general areas that agricultural engineers specialize in. Soils and water is the first and concerns itself with water drainage, erosion control, irrigation systems, and land use. The second area is food engineering, where engineers analyze food-processing procedures, finding ways to minimize waste, energy, and damage. Drying of foods would fall under this category; perhaps if you are a backpacker you have enjoyed the benefits of this technology. In this process the food is placed in a vacuum chamber where the water evaporates; then the food is sealed in an airtight package to prevent spoilage. You can easily imagine how many steps the process takes, all of which must be considered in the engineering design. Agricultural engineers are also involved with irradiation of food as a means of long-term safe storage. A third area is power machinery, the development of new agricultural feed systems and handling and processing machinery. The fourth area deals with structures, particularly those necessary for the housing of livestock and the food and waste-handling aspects necessary for this type of building. Last is the area of electric power utilization. In remote areas farmers and ranchers must generate their own electricity. The wise and efficient use of electricity is always important, particularly so in self-reliant situations.

Architectural Engineering

The architectural engineer works with architects in the design of buildings, focusing on the analysis and design of materials used in construction. The architect in this situation is concerned with aesthetic requirements of the space and the environment it surrounds and surrounding it, while the architectural engineer's concerns focus on the structural integrity and safety of the design. There is a great deal of similarity between this and structural engineering, the main difference lying in the architectural engineer's concern for aesthetic considerations.

Automotive Engineering

One of the major industrial forces in the economy of the United States is the automobile industry, and engineers are a vital component of its growth and strength. Automotive engineering transcends automobiles and includes all types of vehicles from trucks, to bulldozers, to motorcycles. The field is an interdisciplinary one, as you can readily imagine by envisioning all the systems in today's automobile. There is, of course, the design of the engine itself, with its thermal and mechanical aspects, as well as the fuel and lubricant considerations. The structural design must be such as to withstand impact and protect the driver and passengers. Here too the structural design depends on the materials used, with thermoplastics replacing metal in many instances, and on the aesthetic requirements so fundamental to marketing the car.

Automotive engineers use workstations to gather and analyze crash data. Strain gages are attached to the car's chassis and connected to the workstation. The computer is so fast that it can measure all the strains virtually simultaneously during a crash. Engineers use the data to determine where to make design improvements, creating a safer automobile. (Courtesy of Hewlett-Packard)

Biomedical Engineering

Biomedical engineering is a comparatively new field of engineering that began in the 1940s but, in a sense, has been practiced for thousands of years since the creation of the first artificial limb. The field is interdisciplinary, and engineers must be able to work with biologists and physicians in developing new equipment and materials. This means being able to communicate in their language as well as in the language of the engineering community. The three general divisions of biomedical engineering have names that are frequently used interchangeably. Bioengineering, a research activity, applies engineering techniques to biological systems. Medical engineering develops medical instrumentation, artificial organs, prosthetic devices, and materials. Clinical engineering concerns itself with the hospital system problems, such as decontaminating air lines, removing anesthetic gases from operating rooms, and the correct operation of instrumentation in health facilities. You would certainly want the monitor that a physician was using on a patient to accurately reflect the condition of the person. Very often this field requires a graduate degree because of the diverse nature of the material that must be comprehended.

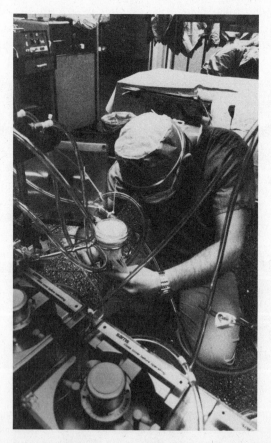

A technologist is installing a filter in a cardiopulmonary bypass circuit—a system design by bioengineers. (Courtesy of Pall Corporations)

Ceramic Engineering

When we think of ceramics, the image conjured up may be a green frog, a bowl, or a pitcher. Whereas these objects are called ceramics, the field of ceramic engineering is indeed something far different and very interesting. Ceramic materials are nonmetallic, inorganic materials, such as silicon dioxide (a form of sand), that fuse at high temperatures to form a variety of materials. The space shuttle had much of its surface covered with ceramic tiles to withstand the high temperatures of reentry into the earth's atmosphere. Ceramic materials can withstand high temperatures and not lose their strength and, because they are chemically stable, can be used in corrosive and chemically active situations without deterioration. Rocket nozzles are made of ceramic materials, as are spark plugs. Ceramic engineers are needed to determine the material requirements for the particular application. For instance, diesel and gasoline engines have some of their metal surfaces coated with a thin layer of ceramic material to improve efficiency. The piston crown and exhaust valve seats are so coated and therefore can withstand a greater temperature. Also, the engine loses less heat to the cooling water, due to the insulating aspects of the ceramic coating; this allows the engine to convert more of the fuel's chemical energy into work. A ceramic engineer had to design the ceramic material to withstand the vibration of the engine and the thermal expansion of the metal in this situation. So ceramic materials do have some flexibility, even though we think of them as brittle.

Ceramic engineers were fundamental to the development of fiber optics. This engineer is operating a fiber optic test station. (Courtesy of Hewlett-Packard)

Chemical Engineering

Chemical engineers translate the laboratory developments of the chemist and physicist into commercial realities. The chemical engineer works in the pharmaceutical, chemical, pollution control, nuclear, and electronics industries. In the electronics industry, for instance, chemical engineers must confront and solve the production problems encountered in the manufacturing process. The most commonly used method to manufacture electronic components is evaporating and redepositing a material on a surface a distance from it in a vacuum chamber. The substance is housed in a vacuum chamber and has a large amount of electrical current passed through it. The substance evaporates, with its atoms traveling to a surface about one foot away and deposited there to a thickness of less than one micron. Remember that a micron is one-millionth of a meter or one thousandth of a millimeter. This process must be accurately controlled and measured—a job for a chemical engineer.

More in line with what we imagine a chemical engineer doing, in the pharmaceutical industry a new antibiotic, Xmycin, has been developed by the biochemists in the research laboratories. The next stage is to design a pilot plant to manufacture Xmycin. Problems of scaling up the processes must be analyzed and overcome, and the equipment must be designed and controlled. Once the manufacture of the antibiotic is successfully accomplished in the pilot plant

These electronic circuit boards are cleaned with extremely pure water. This technology, developed by chemical engineers, is environmentally preferable to one using fluorinated hydrocarbons. (Courtesy of Pall Corporation)

setting, the company can decide whether or not to proceed to a full-scale manufacturing facility. Chemical engineers must redesign the equipment for this situation.

Chemical engineers are involved with all types of processing equipment and manufacturing, from petroleum products, to paints, to nuclear fuels, to vitamins. All of these plants have chemical processes to control and equipment that must be designed for the characteristics of the substances under consideration.

Civil Engineering

The field of civil engineering is one of the most apparent to us. The highways and bridges we drive on, the tunnels we drive through, the buildings we live and work in, and the water supply and sewage treatment systems we depend on are the handiwork of civil engineers. As civil engineers you need to understand a variety of situations that go beyond engineering mechanics. To be sure, part of your course of study and career opportunities lie in the construction industry, structural design, and overseeing the construction process. The building rests on the ground, and you must understand and be able to determine the load-bearing properties of soil, and for instance, how to increase the load properties with the use of pilings.

Fundamental to the design of structures and roadways by civil engineers is a land survey. (Courtesy of Sidney B. Bowne and Son)

The construction of highways requires accurate surveying of the terrain, showing exactly where the roadway is going and defining aspects of the terrain that need to be considered in the highway construction. A highway is more than the construction of the roadbed: the engineer must consider the effect of drainage, land use and environmental impact, and the interconnectedness with other transportation systems. Bridges are one of the engineering achievements that most strongly capture our imagination, such as the Golden Gate Bridge in San Francisco or the Verrazano Bridge in New York City.

Another important area of civil engineering is waste treatment, including garbage, sewage, heavy metals, and toxic materials. We personally create a tremendous amount of waste in the course of our daily lives. In 1985 the household waste generated in the United States was 136 million metric tons. (A metric ton is 2200 pounds.) This does not include waste from the industrial sector. How to dispose of this is one of the major problems facing municipalities today, and civil engineers can help design the necessary systems.

Because what civil engineers construct is so vital to society, most civil engineers are licensed professional engineers, more so than engineers in any other engineering field. Also, because of their involvement with large-scale municipal systems, many of the engineers employed by local and state governments are civil engineers. Over 50 percent of the city planners in the United States have a background in civil engineering.

Most surveys require line-of-sight measurements. However, a global positioning system allows engineers to determine a three-dimensional position on the earth's surface using satellite communication. This highly accurate method can be used effectively in urban areas, even where buildings block the line of sight, and in geographically difficult regions, such as hilly terrain and canyons. (Courtesy of Sidney B. Bowne and Son)

Computer Engineering and Computer Science

Computer engineering is closely connected to computer science, and while most of computer science is not an engineering field, students often ask whether they should study computer engineering or computer science. Both are involved with the design and organization of computers: its software and hardware. If we view this amalgam being the area of commonalty, computer science approaches this from the software viewpoint with less emphasis on the hardware, and computer engineering from the hardware viewpoint with less emphasis on the software. Computer science in the most general sense is the study of problem-solving procedures, computability, and computation systems. Calculus, which is based on the mathematical differential, is not as relevant to computer science as finite mathematics, because digital computers deal with discrete pieces of information. The field of numerical analysis concerns itself with approximate solutions of sets of equations on digital computers. Software engineering deals with the creation of software to solve problems.

Electrical and Electronics Engineering

Electronics engineering is often called electrical engineering: however, electronics engineering excludes large electrical systems, such as found in motors, generators, electrical circuits of buildings, and power transmission systems. Electronics engineering deals with the passage of charged particles in a gas,

Electrical engineers are found in urban, suburban, and rural environments. This engineer is carrying a portable spectrum analyzer to check a remote telecommunication site. (Courtesy of Hewlett-Packard)

vacuum, or semiconductor. Electrical engineering, exclusive of electronics, deals with the motion of electrons in metals, such as through wires and filaments.

This description does little to show the tremendous variety of ways in which electrical engineers are involved in today's society. A blackout dramatizes how dependent we are on systems with electrical components. Electric power eases our daily life tremendously. The typical U.S. household has at least ten electric motors, and this number can easily be multiplied several times for some homes. Every fan, dishwasher, washing machine, and refrigerator relies on motors for its operation. Automobiles are also equipped with a variety of electrical and electronic devices, from the fundamental battery to sophisticated electronic fuel injection systems. All are designed by teams that include electrical engineers.

The growth of electrical engineering in the last 40 years has been exponential. Microprocessor-based control systems are used in most major home appliances, such as microwave ovens and dishwashers. Computer-based control systems are standard equipment in aerospace systems. Control systems provide rudimentary intelligence to devices, so they can sequence operations. Some modern aircraft are unstable unless computer-controlled; pilots cannot respond quickly enough to fly forward swept wing planes without computer assistance.

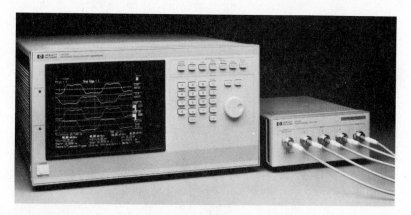

Electrical engineers created this multichannel digitizing oscilloscope. The oscilloscope digitizes the analog signal, displays the signal, and can store it or send it to a computer. (Courtesy of Hewlett-Packard)

In experimental activities instrumentation frequently relies on an electrical (electronic) signal to show that a certain event has occurred. For instance, thermocouples measure in millivolts a voltage potential that correlates directly with temperature: knowing the voltage allows determination of the temperature.

It is difficult to imagine an engineering discipline that does not include or require the use of electrical engineering at some point in the system design. Let's consider a lesser known area of electrical engineering, oceanic engineering. Electrical engineers are developing technology and analysis methods to determine the sea state of the ocean through satellite observations. The satellite emits and receives an electrical signal that is distorted by the ocean's sea state, the wave height and velocity, and wind-generated surface conditions. By knowing what the weather conditions are in remote ocean areas, ships can avoid turbulent weather, and weather forecasts can be more accurate.

We are in what some people call the communication age, the age of information overflow. How is this information created and transmitted? By electrical and electronic devices. Computers generate information from programs they run; the information may be transmitted across country by satellite and through optical fibers, or stored on diskettes.

Every piece of electronic or electrical equipment requires the knowledge that an electrical engineer possesses to be designed, developed, and manufactured.

Environmental Engineering

This specialization in engineering has traditionally been affiliated with the civil engineering programs at most universities. It is, however, developing a separate identity and crosses the boundaries of many disciplines. All engineers are concerned about the environment and in creating processes and products that minimally disrupt the natural environment. Thus, environmental engineers may be chemical engineers focusing on the containment of environmentally hazardous materials within a plant, preventing their release. They may be mechanical engineers concerned about air pollution caused by combustion processes, or civil engineers looking at waste disposal problems and water quality issues. Environmental assessment of manufacturing plants, proposed and existing, is very important in the public eye, and engineers are creating plants, facilities, and systems that minimize adverse environmental effects during normal operation and in cases of natural disasters.

Industrial Engineering

Industrial engineers are concerned with the design, improvement, and installation of integrated systems of people, materials, and energy. What are examples of these systems? Not all systems have to have all components. Consider the manufacture of tires: a machine is operated by a person and needs certain raw materials to produce the tire. The problem for the industrial engineer is to bring the material to the machine as it is needed, remove the finished product, and transport it to the next workstation. When establishing a manufacturing procedure, the industrial engineer considers the machine's tire capacity, when the machine needs routine maintenance, and what the machine operator should do when the machine is not functioning. Industrial engineers are always analyzing manufacturing processes to determine if the sequence of operations can be made more efficient so the manufacturing process is less costly per tire, making the company more competitive, yet maintains or improves the tire quality. The role of the industrial engineer has increased and is critical to a company's financial success, as the technologies that must be integrated are

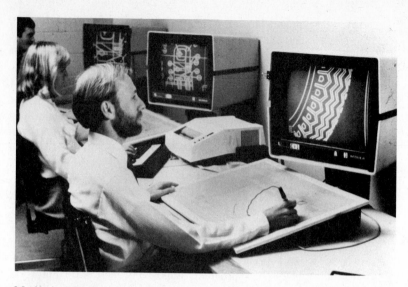

Mechanical engineers designing a tire mold. (Courtesy of Michelin Corporation)

increasingly complicated and costly, and the competition in the world marketplace is great.

Your background as an industrial engineer will include areas in management, manufacturing, plant design, quality control, data processing, and systems analysis. Industrial engineers must be people-oriented and will find themselves involved with incentive programs, employee benefits, subcontracting and contracting, and areas of financial management.

Manufacturing Engineering

Manufacturing engineers are concerned with producing a product at the lowest possible cost, in the shortest time, while meeting all its design requirements. The manufacturing engineer is well aware of the budgetary constraints of management in producing the product and of the desires of the customers who want the product they purchase to be as described by sales and marketing personnel.

When a decision is made to produce a product, the manufacturing engineer plans what facilities can be used and the sequencing of the production operations. If the objectives set for the product manufacture cannot be met by the equipment on hand, the manufacturing engineer specifies what type of equipment will be needed. However, manufacturing engineers get involved long before the end of the product development process.

They work with sales and marketing on an ongoing basis, adjusting the capabilities of the manufacturing plant in anticipation of new products and the time period needed for production. Manufacturing engineers also work with design engineers in determining product specifications based on current and/or projected machinery availability. They may evaluate the current overall manufacturing performance and make recommendations regarding improvements to the system. When new equipment is purchased, they assure its integration and compatibility with existing machinery. A manufacturing engineer may also be involved in the purchase of nonproducing facilities and equipment such as buildings and vehicles, judging the compatibility aspects of these acquisitions.

Marine Engineering, Naval Architecture, and Ocean Engineering

The oceans form 80 percent of the earth's surface, and transportation of goods across them and the extraction of resources from them is the domain of these engineering specialists. The naval architect designs the ship's structure; its hull form and the interaction between the hull and the water are of paramount consideration. The design must consider the control of the ship and its stability under all conditions. Since ships are floating cities or hotels, all ventilation, water, and sanitary systems fall under the naval architect's purview, as well as the navigational systems and the propulsion system. Naval architects work with marine engineers, who are primarily responsible for the design of the ship's propulsion and auxiliary systems as well as the automatic control systems governing them.

The offshore oil drilling rig must be able to withstand hurricanes as well as the highly corrosive saltwater environment, a challenge to ocean engineers who designed it. (Courtesy of ASME)

Ocean engineers design the vehicles and devices that cannot be called a ship or a boat, such as offshore drilling rigs, offshore harbor facilities, and underwater machinery. For example, this type of machinery is used for harvesting minerals from the ocean floor.

Materials and Metallurgical Engineering

Materials science seeks to understand the properties of materials, the fundamental bases of their behavior. What is there about the atomic and molecular structure that will make one material brittle, another elastic? Materials engineering uses this knowledge to develop new materials that will have improved characteristics—greater strength, corrosion resistance, fatigue resistance. With the new materials, new designs can be created and new manufacturing processes developed. This explains, in part, the dynamic nature of engineering, ever growing and changing. All areas of industrial production, from automobiles to hearing aids, are affected by gains in materials science.

Metallurgical engineering is the area of materials engineering that is concerned with metals. Extracting metals from naturally occurring ores— making steel from iron ore, aluminum from bauxite—is one domain of the metallurgist. Another is the development of alloys, starting with bronze in 3000 B.C.

Mechanical Engineering

Mechanical engineers apply the principles of mechanics and energy to the design of machines and devices. Often we associate mechanical engineering design with devices that move, such as a lawnmower or food processor, but it includes thermal design as well, such as air conditioning systems. The mechanical engineer must be able to control the mechanical systems and frequently works with electrical engineers in designing these systems.

Applied mechanics is the study of motion and the effect of external forces on this motion, so the mechanical engineer is involved with engine crankshaft design and turbine rotor design. Engineers in this area must consider the vibration the device causes on the system and the counter situation of the vibrations imposed on the device. In designing the nozzle of a rocket the engineer also looks at the design from two viewpoints, the fluid's effect on the nozzle and the nozzle's effect on the fluid. There is a keen interest in the interactions between the fluid and solid interface, such as in the imparting of fluid energy to turbine blades to produce power.

Whenever there is motion, there is wear of the moving materials and eventually the piece will break. To minimize wear mechanical engineers must understand lubrication, choosing the lubricant that best inhibits wear between surfaces in relative motion. This is not an easy task: the lubricant must be able to withstand the operating environment of the machine and be contained within a particular space and not contaminate other regions of the device. As the mechanical engineer designs a new machine or device, he or she must be aware of new materials and their selection for gears, brakes, housings.

The field of mechanical engineering is the second largest engineering field, following that of electrical engineering.

Mechanical engineers design objects small and large. This massive crankshaft for a large diesel requires the same precision as one thousands of times smaller. (Courtesy of MAN B&W)

Mining and Geological Engineering

Mining and geological engineering are closely allied fields involved with the discovery and removal of metals and minerals on earth. The ore bodies may be located on the earth's surface, underground, or under the ocean floor. The geological engineer is, in general, more concerned with exploration and mapping of ore bodies than the mining engineer, who must deal with the systems necessary to extract the ore. Before a given mining method can be selected, the size of mineral deposit must be estimated by test drillings.

For surface mining, such as strip mining and quarrying, the engineer must have knowledge of soils and rocks, blasting techniques, resource management skills, and, importantly, environmental restoration. The engineer will devise plans to return the area to a natural state after the mineral deposit is removed. In underground mining the engineer must be cognizant of mining techniques, including safety considerations, ventilation requirements, draining and pumping of tunnels, surveying, and boring and blasting methods. Mining engineers are also involved with extraction of minerals from the ocean floor and below the floor.

Nuclear Engineering

The word *nuclear* conjures up visions of disasters, from bombs to the power plant failures at Three Mile Island and Chernobyl, and to be sure the dangers of nuclear fission and fusion are real. However, the success in engineering design is overwhelmingly positive, essentially 100 percent, as operator errors have been the cause of the aforementioned calamities. Nuclear engineers also design, develop, and build systems other than power plants. For instance, radiation is used extensively in medicine, particularly in combating cancer. Irradiation of food has proven to be an effective and safe method for long-term storage of bulk commodities such as grain and of highly perishable ones such as milk and bacon.

The electrical power generated by the 92 nuclear power plants in the United States was 79,000 megawatts in 1986; France produced 39,000 megawatts with 44 power plants, and Japan 24,000 megawatts with 33 power plants. These plants were designed by nuclear engineers, and the control of waste and future developments must be devised by these engineers. Materials were developed to withstand high levels of radiation. A spin-off technology from nuclear engineering is radiation techniques that detect hidden flaws in materials. The problems of fission plant design and fusion plant design and development remain

Nuclear power plants generate a significant amount of electricity in the United States (16 percent of the total) and in France (65 percent of the total) as of mid-1986. The problem of safe nuclear waste disposal is a cause for concern, and nuclear engineers are seeking ways for secure disposal. (Courtesy of Public Service Electric and Gas Company)

for nuclear engineers to solve. One of their most significant and challenging problems is the disposal of nuclear waste and the reprocessing of radioactive materials.

Petroleum Engineering

Petroleum and petroleum products are the lifeblood of our industrial society, and the petroleum engineer is involved with sustaining its flow. Petroleum engineers are involved in all stages of the oil and gas discovery. Working with geologists and geophysicists in locating the petroleum sites, engineers use techniques such as seismic mapping to denote the areas to be drilled. Petroleum engineers are involved in the design and use of drilling rigs, offshore platforms, and drilling equipment. These platforms are often located in hundreds of feet of water, and the drilling must extend thousands of feet beneath the ocean floor.

When a site is found, the oil or gas must not be lost to the environment and must be removed from its natural state to storage and transportation facilities. In the primary recovery phrase, the oil or gas flows under its own pressure to the surface and can be pumped. The secondary recovery phase occurs when petroleum engineers use techniques such as water flooding, gas injection, or in situ combustion to force the petroleum to the surface. Certainly the pipelines in the storage areas require the knowledge of a petroleum engineer for their design, as do the natural gas pipelines crossing the country. One of the most challenging engineering projects for petroleum engineers, working in concert with engineers from other disciplines, was the development of the Alaskan pipeline, which has to transport heated crude oil in an environmentally sound manner across hundreds of miles of Alaskan tundra.

3.5 TECHNICAL CAREERS IN ENGINEERING

Not all engineers have the same work function; the variety is quite large, depending on the type of industry or business as well as the company's size. Thus far we have looked at the types of design projects in which you might be involved within each of the specializations, but this is only one part of the panorama of opportunities that awaits you.

There are seven general areas of activity for engineers involved with a project, such as building a dam, or manufacturing a product, such as a refrigerator or VCR. These are research, development, design, manufacturing or construction, operations and maintenance, sales, and management. Some of the functions will be combined in companies. In addition the area of quality control and quality assurance is becoming very important in all areas of engineering, but predominantly in the design and manufacturing areas. In the course of your engineering career you will often be involved with several of these areas before selecting, in most cases, one specialization. Let's look at what is involved in each area, remembering that this is superimposed on a given field of engineering.

An engineer may often make use of a workstation such as this one. (Courtesy of Hewlett-Packard)

Research

Research in an engineering setting is not the basic research of scientists seeking to explain phenomena not understood, but rather applied research that uses the results of basic research. The development of ceramic tiles for the space shuttle or of composite materials for aircraft wings is applied research.

Why bother in the first place? In general, research is aimed at creating a new or improved product. Applied research has resulted in the creation of thermoplastics, which can be used to replace some metal parts and strengthen others in automobiles. Research creates a new technology, market conditions determine whether a new product is needed, and, presuming it is, development engineers create the product. Before we address the development function, what types of engineers are usually found in the research areas of companies? Very bright people with advanced degrees, often doctorates, who like the challenge of working in poorly defined areas of research and who exhibit high levels of creativity. Of course open-mindedness is a requirement, so that all possibilities percolate to the surface and are not discarded prematurely. Just such an attitude was required and exhibited in breakthrough developments in superconductivity.

Development

Development engineers work closely with research engineers; in many organizations these functions are called R & D, because companies cannot afford

A research engineer using electrospectroscopy to detect contaminants as small as one atomic layer thick. (Courtesy of United Technologies)

undirected research. The direction must be towards product improvement or new market potential. Indeed, the development engineer must convince management to fund various new proposals based on market projections or customer requests from the sales engineer. In chemical engineering development might mean the construction of a pilot plant, in aerospace engineering, the testing of scale models in a wind tunnel. The development engineer does not always deal with the latest research, but may use existing devices in a novel way to solve a problem. When there is a problem and a new product is needed, the development engineer will create several alternative solutions.

Design

The next step in the product or process development is taken by the design engineer. After the development engineer shows that a new product or process is possible and management is convinced to expend its resources in people, space, and money to produce it, the design engineer takes the model or concept and translates it into a producible form. For a product, the development engineer will propose several models to show that it can meet the goals established for it, such as improved efficiency or use of microprocessor control. The design engineer selects and modifies the product in light of the manufacturing facilities available: you cannot specify a tolerance smaller than the manufacturing machines can produce. The design of the product must be produced in such detail that engineers in manufacturing, though unfamiliar with the product or even its purpose, can set up the machinery, the material specifications, and the manufacturing components required to manufacture the product. The same

Not all engineering design work is done on a computer. This engineer is completing the design of a metallic filter at a drafting table. (Courtesy of Pall Corporation)

scenario is involved for a process, such as a new paper-making technique, or for a one-of-a-kind product, such as a dam or large office building.

In designing the product the design engineer must constantly balance economics and product quality. Obviously, you want the product to be of outstanding quality, but you must bear in mind that all materials wear out with time. Consider the automobile engine: most engines will last over 100,000 miles with modest maintenance, but some components, such as filters and spark plugs, must be replaced periodically. It makes little economic sense, in light of the competition, to design an oil filter that would last 100,000 miles, as it would be very expensive. The competition's filters selling for $2 to $4 would continue to be chosen by consumers.

While on the topic of automobiles, a designer must consider the trade-offs involved between using normal-gauge sheet metal, thin sheet metal with thermoplastic covering for additional strength and rust resistance, or fiberglass, all while considering the final shape of the body, its aesthetic value, and the manufacturing costs associated with each. A design engineer must have a broad engineering background and work well with the often conflicting objectives of cost, quality, and aesthetics.

Manufacturing and Construction

Manufacturing and construction engineers face a similar challenge—to translate the paper renderings of a product or building into physical reality. This takes the coordination of many types of people, materials, and equipment. Consider that a tire company wishes to manufacture a new tire. The equipment required may be some existing machine and perhaps a new machine to perform a special function. The assembly line must be established in conjunction with existing products, the new machine ordered, materials ordered, and personnel coordinated. The manufacturing engineer must be able to communicate and work well with trades people, design engineers, and management. This person may have great influence on the product quality and cost, and hence the long-term viability of the company.

The same abilities the manufacturing engineer possesses are necessary for the construction engineer if a bridge or other construction is to be completed on time and on budget. The engineer must be able to handle labor disputes, quickly communicate to management significant problems that affect project completion, and assure a flow of materials to keep the work force fully utilized.

Operations and Maintenance

Once any facility exists, be it a manufacturing plant or a university, it must be maintained so that the buildings and the manufacturing machinery operate as they are designed to do. The operations engineer must be a generalist and be able to deal with electrical systems, electronics, control systems, and steam systems, and the specialists who repair and maintain them. The technicians in each area

A construction engineer must coordinate the activities of many trades to keep a project on time and on budget. (Courtesy of American REF-FUEL)

will maintain the equipment; they must be scheduled, and the schedule is affected by the analysis of the operating data from the equipment. The analysis of this data is part of operations engineer's responsibility. Do the data indicate that the circulating water pump for the air conditioning is failing and needs to be replaced? How soon it will fail and when to assign someone to replace it without disrupting a building's work force are questions the operations engineer must answer. Computer systems require a cool environment, and businesses such as financial institutions would be in disarray if the air conditioning failed, all for lack of a pump replacement. It is with such pressure and with such awareness that the operations engineer must assess work assignments. Operations engineers are also responsible for plant safety and technician training, and must be knowledgeable about union contracts and the resulting job jurisdiction.

Sales

A vital element in the survival of any company is that its product be purchased by others. Having a superior product or service is not sufficient; someone has to make it known to consumers. Sales work in engineering is different from retail sales, which is what we encounter in department stores. The sales engineer must be knowledgeable about the needs of the customer and why a product will satisfy the customer's requirements and, in all probability, will do so better than the present system or a competitor's product. The sales engineer must be a technical generalist able to deal with a variety of customers and at ease with communicating with them: this means being able to listen as well as talk. Very often it is the sales engineer who lets the company know about new product needs, and this information flows to the development engineer. The sales engineer must be technically proficient, so that she or he can relate what the advantages of a new product are to other technical people and answer their questions in technical and general terms.

Additionally, the sales engineer is the visible representative of the company, that others see, that they form an opinion of the company, and thus companies are selective about whom they choose for sales. It is seldom an entry-level position because of this and because it takes time to learn the details of the product line. Usually a sales engineer starts in the manufacturing or service area, to gain a generalist background regarding the company.

Management

Many engineering students do not want to pursue a lifelong career in technical engineering but wish to move eventually into management. Studies have indicated that upwards of 75 percent of engineers become managers. Part of the reason is the societal recognition that such promotions lead to large salary increases for those at the top.

Such notables as Lester C. Thurow, dean of MIT's Sloan School of Management, have pointed out that when it comes to inventing new technologies, the United States has no peer. Using the technologies in the industrial arena is where we lack. The managers of today are reluctant to use new technology, and by the time they adopt it, foreign competition has several years' head start. Engineers need to be involved with more than inventing the technologies; they should move into upper management to introduce technological innovations more rapidly into industry. Nontechnical managers certainly can comprehend new technology, but they do not have an intuitive understanding of which option to pursue. Engineers, because of their education and training, have developed this intuition and can learn the management skills necessary to control technical industries. It may not be sufficient in the future, as is indicated by industry's present state, to have managers in technical industries who have no technical background.

What do managers do? They manage resources—people, money, and materials. They do not create, but can be creative in the management of these resources, accomplishing a given task or implementing a phase of corporate philosophy. We all have to be managers in our daily lives, so the concept is one we are familiar with; we plan our individual and family lives and actuate the plan as best we can. We manage. However, when you are directing not just yourself but others, you need special interpersonal skills and aptitudes.

What do you do as a manager? Attend meetings, read reports and memoranda, and write reports and memoranda in support of getting things done with and through others. Thus, you must be able to plan, organize, direct, and control several tasks or activities simultaneously in support of a given objective. The objective might be to design the landing gear on a new aircraft, or to develop the specifications for a circuit board. As a manager you are responsible for and rewarded for the work that others perform.

One of the significant advantages of your engineering education and of the aptitudes you must possess to succeed in engineering is that you have many of the qualities necessary to be a good manager. Most management situations involve solving a problem, be it producing a new dishwasher or controlling vibrations in a space station. Your education is directed towards solving problems, problems in the physical world. Many of your managerial difficulties will be in solving people problems, a skill that can be learned and an art that can be refined.

Government, Consulting, and Teaching

Government, consulting, and teaching are the other areas where you might seek a career. Many engineers work for federal, state, or local governments in a variety of capacities. Our governments are extremely complex, exist in a technological world, and need engineers as pathfinders. Engineers provide an important link in allowing government officials to make sound decisions in light of complex technological systems. Engineers help resolve environmental questions in most public construction projects, oversee the construction and maintenance of public facilities, and direct public transportation systems and public utilities (electricity, water, and waste). In addition the government funds fundamental and applied research in the development of new technology and employs engineers to oversee or undertake such projects.

Consulting and teaching are two smaller areas of practice you might consider as career paths. Often companies find it more cost-effective to hire specialists than to keep them on staff. Thus, consulting firms originate with specialists who decide to work on their own and who have recognized expertise that they can market. You can do the same, but must develop the expertise by working in industry in a technical area. More frequently you can work directly for a consulting engineering firm that will train you as you work in their specialty areas.

Teaching is another area that some engineers pursue. An advanced degree, most often a Ph.D., is required, and a desire to work with students and to conduct research should be an attribute. You must be able to balance the research and teaching interests and have an independent mind to benefit from this environment.

Quality Assurance and Quality Control

Quality control and quality assurance are often part of the manufacturing process, but in many organizations they are a separate department. Because of

Quality control engineers assure that the product produced meets the company's specifications. This engineer is checking the tire seat on the wheel rim after a test. (Courtesy of Michelin Corporation)

the special emphasis quality control and assurance are receiving in the United States currently, they will become a more apparent career path within organizations. With increased competitiveness in the world marketplace, any product or service must be designed and manufactured with quality control checks to assure the required performance is achieved.

Quality control means checking materials and components as they are received and manufactured, so that the steel has the correct carbon content, and the size specifications for certain materials are within tolerances. Quality assurance means checking the product or service after it is completed to assure that it meets the initial objectives. The combination of both aspects creates the quality system that is increasingly used in companies. We have all been made aware of the need to improve the product quality of many U.S. manufactured goods over the past decade, by articles in the popular press and professional journals. Companies are responding and creating stronger and more active quality assurance/control departments. They have rediscovered that quality control means greater profitability for the company.

3.6 WHAT IS QUALITY?

Let's describe some aspects of quality, at least in regard to its use in quality systems. In the quality of a product or service, we expect a greater degree of excellence if *the price* is high compared to a competitor's, if *the cost* to produce

Continual testing is one element in assuring product quality. This tire is undergoing a laboratory cornering test to the point of failure. Analysis of the failed tire will yield information necessary for engineers to create design improvements. (Courtesy of Michelin Corporation)

it is higher, and if it is less *variable* in performance from piece to piece. The degree of excellence refers to the performance and advantages of the product prototype vis-à-vis competing products. Once the prototype is created, quality control becomes necessary, as manufactured products will vary from the prototype even under the best of circumstances. For instance, the prototype may assume a soldering technique that is difficult to perform, creating a high incidence of defects. The design should be revised in light of this information. This is the design review stage. So not only manufacturing is involved, to check that the various steps can be accomplished with the facilities available, but also quality control, to provide input regarding the repeatability and accuracy of the steps. To control the variability of the product the design must allow for the variations of performance of the raw materials and components. The engineers in quality control have knowledge about this variability in other situations and can help determine if it can cause problems in the product performance. For instance, an electric circuit uses a capacitor of a certain size, let's say 10 microfarads. A quality control engineer knows that the shipments of capacitors of this size will vary from 8 to 12 microfarads, so the design must allow for this variation. Otherwise the product may not perform satisfactorily, and the rate of returns will be too high, resulting in decreased profitability or even losses.

There are costs associated with implementing a quality control system. Appraisal cost includes the actual costs of making the quality assurance tests on the completed product and the quality control tests on the raw materials and components, as well as the cost of the design review stage. Failure costs result from a product not meeting its performance specifications. These costs arise from the remanufacture and repair of products that have been returned because of substandard quality. Accidents that occur because of substandard product performance and warranty reimbursements are both in this category. A determination made in the quality assessment of a new product, that additional training of manufacturing or shipping personnel is required, adds to preventive costs, expenses made to reduce the failure and appraisal costs.

These three categories can be combined to form a chart (Figure 3.1), depicting the total quality cost. Notice that the minimum of total quality cost assumes there will be A percent of defective products and that the quality program will incur B of expense. To spend more effort and expense on prevention and appraisal decreases profitability, as does spending less and incurring more failure costs as a result. Inherent to this model is the maxim that product quality must be attained with limited financial resources and that the product must have a market that will return the investment. Engineers realize that companies must produce products that are competitive in price and better in performance than others on the market. This is a fundamental challenge in designing a new product. Quality begins with the creativity and producibility of the design and follows through the manufacture and delivery of the product to the customer.

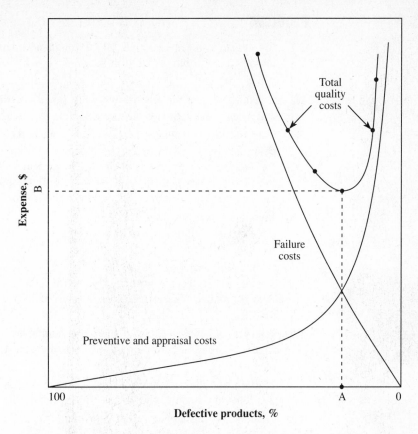

FIGURE 3.1 A total quality cost diagram formed by the superposition of failure costs and preventive and appraisal costs.

REFERENCES

1. Badawy, M. K. *Developing Managerial Skills in Engineers and Scientists*. Van Nostrand Reinhold, New York, 1982.
2. Beakley, G. C.; Evans, D. L.; and Keats, J. B. *Engineering: An Introduction to a Creative Profession*. 5th ed. Macmillan, New York, 1986.
3. Carruba, E. R., and Gordon, R. D. *Product Assurance Principles*. McGraw-Hill, New York, 1988.
4. Jamieson, A. *Introduction to Quality Control*. Reston, 1982.
5. Kemper, J. D. *Engineers and Their Profession*. 3rd ed. Holt, Rinehart and Winston, New York, 1982.
6. _____. *Introduction to the Engineering Profession*. Holt, Rinehart and Winston, New York, 1985.
7. Waterman, R. H. *The Renewal Factor*. Bantam, New York, 1987.

PROBLEMS

3.1 Investigate a recent research breakthrough and determine some of the important steps that occurred in the process.

3.2 What is the difference between development and design?

3.3 Consider each of the engineering work functions from research to sales, and discuss the aptitudes that are desirable for each.

3.4 Using a national engineering periodical of one of the engineering societies, determine the current and projected demand for engineers in general and for the type of engineer you wish to be in particular. You may also consult the Sunday classified ad sections of large metropolitan newspapers to determine the number of companies looking for various types of engineers.

3.5 Imagine a new product that you would like to produce. Then assign yourself the various roles necessary in the design and development of the product. Consider the questions and decisions you will make in each role, then determine which role seems to fit you best.

3.6 Sometimes a project is more appropriate for certain engineering disciplines than a product. Repeat the process in problem 3.5, but this time for a project.

3.7 Interview an engineer, if you know one, and assess what his or her work functions are (design, report writing, meetings, etc.) and the time devoted to each.

3.8 Explain why the demand for industrial engineers has increased faster than the demand for some other engineering specializations in the past decade.

3.9 Discuss the difference between a chemical engineer and a chemist. Interview such people if possible.

3.10 Write a report on the changing job functions of aerospace and aeronautical engineers in the past decade.

3.11 Discuss the impact of composite materials on the engineering design process in the aerospace and automobile industries.

3.12 What is a thermoplastic material? How are such materials used in the engineering design process in terms of material selection?

3.13 Discuss the impact that ceramics have on creating improved products and the material features of ceramics that make them valuable.

3.14 At the library use a reference, such as *Peterson's Guide to Engineering, Science, and Computer Jobs,* to make a list of ten large and ten medium-size companies needing engineers related to your major. Select five of the large ones and determine the high and low selling price of their stock in the past year.

3.15 For the five companies selected in problem 3.14, look for information about their financial health. Books such as *Standard and Poors Index,* found in the financial or stock market section of a public library, give this information. What does this tell you about the company as a place to work?

3.16 Obtain an organizational chart for a company or corporate entity such as a university and examine the hierarchy. Note the variety of information that must be processed by different levels in the organization.

3.17 Major metropolitan newspapers often have a weekly science section. General interest science magazines or journals will discuss technological changes that are occurring. Make a list of the significant breakthroughs in mathematics, science, and technology that occurred in the past year. Note which breakthroughs will have impact on your field of engineering and hence on the management of companies using your field.

3.18 Examine the types of courses required for a masters degree in business administration. Can you categorize them? Contrast them with your technical courses.

3.19 Discuss why the attributes of a leader require more than technical ability alone.

ETHICS AND PROFESSIONAL RESPONSIBILITY

■ To explore our values for ascertaining right and wrong.

■ To discover the elements of a profession.

■ To investigate moral dilemmas: ones you will not face and those you will.

■ To distinguish between product quality and product safety.

■ To develop a greater understanding of the pressures engineers must withstand in practice.

(Photo courtesy of the United States Navy.)

*O*ne of the cornerstones of the engineering profession is its sense of professional responsibility. An understanding of the engineer's ethical responsibility to society is implicit in being an engineer.

4.1 INTRODUCTION

It may seem odd that an engineering course should concern itself with ethics and morality; after all, we know that we should behave in an ethical fashion and have good moral values. What else is there? A lot—and this chapter introduces some of the problems and guidelines and encourages you to consider taking an introductory course in ethics.

What is ethics, and how does it relate to morality? Ethics is the application of a moral philosophy to standards of behavior. We already have opinions concerning what is right and wrong; these opinions are formed as we grow up and are based on a set of moral values we seldom question. The difficult question of what is right and wrong has been a source of discussion and analysis from the earliest human societies. When you are a practicing engineer, a contributing citizen in society, there will be times when you must face decisions that have no right answer. Realizing now that such situations occur and trying to determine the fundamental underlying issues, you will be prepared, at least intellectually, to confront morally perplexing situations in real life.

In the practice of engineering there are professional guidelines which assist in certain areas. These will be examined later in the chapter.

4.2 WHAT IS A PROFESSION?

To practice engineering you must be licensed by the state; hence you are a professional as are doctors, lawyers, and accountants. But other groups are licensed by the state, barbers and beauticians for instance, so there must be more to defining a profession than licensure. Let's examine some of the underlying characteristics of professions.

The manufacture of these printed circuit boards requires the manufacturing engineer to understand materials, electronics, machinery, and people. (Courtesy of Grumman Aerospace Corporation)

Complex Subject Matter

Professionals have extensive knowledge, gained through specialized education, such that the average person cannot judge who is competent and incompetent. Society does not allow the profession to have a caveat emptor ("let the buyer beware") relationship with the consumer, as he or she cannot judge competence.

Professionals also have special skills, such as knowing where to locate precedents on legal matters, diagnostic guides in medical cases, and methodologies and equations in engineering matters. The skills flow from a body of theoretical knowledge that enables the professional to know when to apply them.

Finally, an important aspect that distinguishes a professional from a technically competent person is an awareness of the society in which the skills are being used. Part of the educational background will be devoted to heightening cultural awareness, which is why engineering programs must have a significant portion of study devoted to the social sciences and humanities.

Certification of Competence

Because society depends on professions and we as lay citizens cannot judge professional competence, the state, the collective we, certifies various professions. There are at least two levels to this certification. The educational program is certified, in the case of engineering programs by ABET. Then the professional must pass a state examination, such as the professional engineer's examination.

Upon satisfactory passing of the exam, the state allows you to practice in your profession and call yourself a professional.

Trustworthiness

Since society cannot judge a professional's competence, it must rely on professionals' having an awareness of their important role in society, on their not being self-serving or taking advantage of their position. Often the profession has a self-monitoring aspect, trying to assure the ethical behavior of its members. To implement this policing aspect, the profession educates the member to put society and the client ahead of the self-interest of compensation. To be sure, compensation is important, but it should not be the sole driving force for the professional. The professional takes on work primarily for the psychic satisfaction of the work, and secondarily for compensation from the work as the identity of the professional merges with his or her work.

Another practical way in which professions govern themselves is adhering to an accepted code of ethics. It serves several purposes, as a reminder of the moral standards behind the profession, as a source of guidance in difficult moral dilemmas, and as a standard for evaluating cases of alleged misconduct. The code of ethics of the National Society of Professional Engineers is reprinted in Appendix 1. Several cases that the NSPE Ethics Board has considered are included in Section 4.4.2.

Professional Organizations

Another aspect of a profession is its unique corporate identity, created by similar professionals joining together to form institutions such as ASCE, ASME, or IEEE. The purpose of these institutions is service first to society and second to its members. Such organizations also help perpetuate the professional culture, developing a sense of unity among the members.

Engineering is a profession if you make it one. Herbert Hoover, the thirty-first president of the United States and an engineer, made the following comments about being an engineer.

"It is a great profession. There is the fascination of watching a figment of the imagination emerge through the aid of science to a plan on paper. Then it moves to realization in stone or metal or energy. Then it brings jobs and homes to men. Then it elevates the standards of living and adds to the comforts of life. That is the engineer's high privilege. . . . Unlike the doctor, his is not a life among the weak. Unlike the soldier, destruction is not his purpose. Unlike the lawyer, quarrels are not his daily bread. To the engineer falls the job of clothing the bare bones of science with life, comfort and hope."

The profession to which you aspire contributes much to the growth of new industry and the development of society, and should be viewed in light of its economic and political impact and importance as well as of its daily function of problem solving.

4.3 MORAL DILEMMAS

A study of ethics can help us focus, define, and refine our own moral values. Most often we analyze ethical problems using our beliefs about a lot of different situations. However, our beliefs may be internally inconsistent and contradictory when we examine them. Different regions of the country and the world have different moral values. The task of philosophical reflection is to find a way to incorporate your moral judgments into a consistent and systematic whole.

The Socratic method is used to analyze positions for their logical weaknesses. Socrates questioned others as to their opinions and presuppositions and their implications, and in the process uncovered their logical weaknesses. This same method is used by trial attorneys in examination of witnesses, and can be used to discuss the cases on the facing page.

The purpose of presenting the following moral dilemmas is for you to try to find the moral principal that justifies the action you would take. See if there are situations where you would use a principal in one case and not in another. What are the differences that would make you change your opinion?

As you decide what you would do and why in these situations, be aware of some pitfalls of ethical analysis. The precepts used in making moral distinctions are vague. Although we may believe it's wrong to tell a lie, most of us use "white lies," a lie nonetheless. Some precepts are presumed to be facts, when they actually have a value judgment associated with them. Often people will make a statement that appears to be informative but really is trivially true, such as "when large numbers of people are out of work, unemployment results." There is a tendency towards rationalization: when a person cheats on an examination, she may say, "Everyone does it sometimes," or "The test was unfair." If, on the other hand, someone else cheats, then she is morally indignant. This leads to another problem, what one morally *should* do and what you *would* do.

Often engineers would like to ignore such moral dilemmas and would rather deal with issues related to problem solving in the physical world. These dilemmas are problems to which there is no right solution, but they are indicative of the awareness engineers must develop if they are to participate fully in society. Being attuned to other than technical areas is part of the obligation of being a professional engineer. The political implications, the problems with no complete solution, only minimization of poor solutions, are aspects of life that engineers must constantly recognize. Engineering education tends to mask these indeterminate aspects, seeking quantifiable ones with analytical solutions.

Engineers most often work as part of a design team, making decisions about a product or service that affects the whole, but also being affected by the work of others. The overwhelming majority of engineers work for a corporation, a consulting firm, or a governmental agency and will sometimes be faced with conflicts among loyalty to their employer, to society, to their conscience, and to their profession.

Case I: The Overcrowded Lifeboat

The time is 100 years ago, and a sailing ship carrying passengers and cargo to Europe hit an iceberg and sank. Fifty people survived and tried to crowd into a lifeboat intended for 20. A storm threatened, and the captain decided that some of the people would have to be cast from the boat and left to drown. The captain judged that only the strongest should remain, as they were needed to row the boat. After days of rowing, the survivors were rescued, and the captain was put on trial for his actions. He justified his action because those who drowned would have done so anyway, and doing nothing and allowing all to perish was a worse course of action. Put yourself in the captain's situation. What would you do? Why? Is it different if you are a member of a jury judging the action?

Case II: A Promise

You promised a friend never to tell anyone about a crime she committed several years ago. You have told no one, but now find that an innocent person has been charged with this crime. You plead with your friend to turn herself in to the authorities, but she refuses. What should you do?

Case III: A Low Grade

You are a professor of engineering, and a student who is barely passing with a D comes to you just before the final examination and tells you that he is applying to business school for the next semester. However, to get into business school he will need a C in your course. The student seems to have been working hard but cannot grasp the material. You are tempted to give him the C, as he does not have an aptitude for engineering, and you would hate to prevent him from switching to another school. On the other hand it seems unfair to your other students. What should you do?

4.4 PRODUCT SAFETY AND PRODUCT QUALITY

Engineers are increasingly aware of product safety when designing new or improved products. Society is holding the manufacturer responsible even if the product is used in a way not intended and where intermediaries have sold the product. This does not mean that a product cannot wear out, but it should do so in a safe fashion. Whereas engineers cannot foresee every circumstance in which a product may be used, they must be much more diligent in assessing how it fails. The company that the engineer works for, not the engineer, has been held to be the liable party, and her or his responsibility to society and to the company requires attentiveness in this area. A gray area exists in that product quality is involved. The company must be able to produce its products at a cost low enough to be competitive with others. To design a product that is of the highest quality and consequently has a high and uncompetititve price may mean that the company will not be able to remain profitable and be forced out of business.

This is where the government is important, as it establishes the common standards for design constraints. Consider the situation in the early 1970s when pollution devices on automobiles were first used. The cars were harder to start and less economical—having poorer gas mileage—than the models of a few years before. However, the environment was improved by having fewer pollutants entering it. A company could not have initiated such a design on its own: no one would have bought the car! The same was true for the introduction of seat belts into cars. Many consumers did not like the extra cost, but once belts were mandated by the government (the collective consumer), manufacturers could include them and not risk a competitive disadvantage. One of the engineer's challenges was to design a seat belt system that is economical and meets or exceeds the government specifications.

Manufacturers cannot move too far ahead of what society expects in terms of product quality and stay in business. With new technologies and materials available, the engineer can often redesign and improve product quality without increasing cost. In today's world marketplace engineers are competing with engineers from all nations. The need to maintain or increase product quality while not increasing, but rather decreasing, product cost, is ever more important.

Engineering has often been in the forefront of consumer protection. ASME stepped in to establish boiler standards in the late 1800s to protect citizens from boiler explosions. In 1918 the American Engineering Standards Committee was established, now known as the American National Standards Institute (ANSI). ANSI, through its working committees made up of representatives from all parties and agencies affected by a particular health or safety standard, develop these standards. They advise local, state, and federal agencies who in turn can adopt ANSI standards as regulations.

A marked increase in product liability legislation requires much greater diligence in the design and manufacture of products than ever before. This means that companies become responsible for products they manufactured that are misused. As a result engineers must design products to high standards and often improved quality, for the company to remain in business. The company may be held liable for a variety of reasons.

> If the product's design is defective, it cannot be safely used for its intended purpose.
>
> The manufacturing process is flawed; for instance, the testing and inspection is deficient.
>
> The labeling is inadequate regarding warnings and proper use of the product.
>
> The packaging of the product allows safety-related damage in the shipping process, or the product parts may be separated, allowing sale in a dangerous form.
>
> Records regarding consumer complaints, sales, and manufacturing and distribution information are not properly maintained.

Certainly this seems like a formidable list of requirements, but not an impossible list. Quality control and assurance in all phases of manufacturing from design to distribution have become more important.

Ethical Problems

The drive to decrease costs may lead to some situations that create ethical problems for you. For example, at the end of each month a certain number of computers, electronic components, or tools are expected to be manufactured, shipped, and very importantly, billed. The bills become an asset in the receivables column of the ledger and balance the expenses incurred in producing the product in the debit column. The same event occurs quarterly as well. Businesses, especially small ones, often require short-term loans to cover the time between shipping their products and receiving payment for the products. Banks want to be assured that sufficient payments are in the pipeline, so to speak, before they will give the company a loan, to be used in part to pay your salary. As an engineer you must be aware of business needs as well as the need to maintain product standards, not letting a product be shipped to meet quota guidelines when it falls short of quality or safety standards. Quality standards are different from safety standards: the marketplace will eventually not purchase the product if it fails to meet the quality level expected of it; however, safety standards affect not just the marketplace but society.

There is virtually no moral dilemma in informing your superior that a problem exists with a product, when it violates the law or creates a safety hazard. Your superior will react positively or negatively to your concern. If he or she agrees, then there is no problem. The dilemma arises when the supervisor disagrees or the company decides not to change what it is doing, because it believes it is following the correct course of action. If you make your complaints known outside the company, there is a high probability that you will be fired for your trouble. Perhaps the company officials are correct; you will have sacrificed a great deal, to no benefit for society, if your interpretation of what violates the law or creates a safety hazard is in error. The NSPE offers some guidelines in this area, most importantly the following: "The engineer should make every effort within the company to have the corrective action taken. If these efforts are of no avail, and after advising the company of his intentions, he should notify the client (customer) and responsible authorities of the facts." (Opinions of the Board of Ethical Review, NSPE, 1965). In the rest of this section we will look at some situations where engineers have been in this bind and see what the reactions have been.

The DC-10 Disaster

On June 12, 1972, an American Airlines DC-10 nearly crashed due to a design deficiency in the rear cargo door. The door had to be secured from the outside by the baggage handlers, and people inside the plane could not check the security of the door. Should the door open, the lower cargo hold would depressurize tremendously, the passenger floor above it would collapse, breaking the hydraulic control lines for the rear engine and the tail wings, and the airplane would likely go out of control and crash.

The American Airlines DC-10 did not crash because of several very fortunate events. The captain had, by chance, practiced on a simulator how to handle the plane if he lost control of the rear engine and rear wings. Plus, the plane was lightly loaded with only 67 passengers. When the door exploded out at 12,000 feet over Ontario, depressurizing the cargo hold and causing loss of control of the rear engines and wings, Captain McCormick correctly and coolly reacted, ascertained what the problem must be, and returned the plane to Detroit.

After an investigation, the door problem was solved by installing a one-inch peephole over the locking pins. As early as 1969, however, engineers had noted the problem and suggested changes in the design. The differential pressure problem is one that has known remedies, for instance installing vents on the floor so the pressure differential could not occur. None of the solutions were implemented. On June 27, 1972, as a result of the near crash of the American Airlines DC-10, director of project engineering Daniel Applegate of Convair, the subcontractor and designer of the DC-10 fuselage, wrote a memorandum to his supervisors, stating in part, "It seems to me inevitable that, in the twenty years ahead of us, DC-10 cargo doors will come open, and I expect this to usually result in the loss of the airplane."

Nothing was done beyond minor changes such as the peephole, because of some unusual financial pressures at play. McDonnell Douglas Corporation was in a precarious financial situation and was counting on beating the competition with the newly designed DC-10; delay would allow a competitor's aircraft, the

Lockheed Tri-Star, to be constructed. The assumption was that the market would not accommodate both planes. Convair, a subcontractor, was reluctant to press for a solution, since it was not clear who would have to pay for the design changes, costing millions of dollars. In response to Applegate's memo the general manager of Convair said that he was sure McDonnell Douglas would interpret the recommendation for change as an admission of error on Convair's part. The matter was dropped at that point.

On March 3, 1974, a Turkish Airlines DC-10 with 346 people on board took off from Paris. It reached an altitude of 12,000 feet, the cargo door burst open, the floor collapsed, and the plane crashed, killing all on board.

Nuclear Reactor Welds

The fate of engineers who persist in challenging management on safety issues outside of the company is not good in terms of job security, however beneficial it is to their mental health and self-esteem.

Carl Houston, a welding supervisor for Stone and Webster, a large engineering consulting firm, reported for work at the site of a nuclear power plant being built for Virginia Electric and Power Company (VEPCO) in early 1970. Having no specific assignment given to him, he was free to inspect various welding operations. He immediately found cause for concern. The steel pipes designed to carry reactor cooling water had welds that were substandard for a

Robots perform tasks in environments that are dangerous for humans. This robot is used for routine inspection, monitoring, and surveillance tasks within areas of a nuclear power plant that are subjected to radiation. (Courtesy of Public Service Electric and Gas Company)

variety of reasons. Improper electrodes were being used at times; some electrodes required oven drying they did not receive; and not all of the welders were properly qualified—indeed many were learning on the job. When Houston reported this to the manager, nothing was done to correct the matter, and he was told to take the matter no further. He did not take this advice and reported the matter to the head office of Stone and Webster, which again had no effect on remedying the problem. Carl Houston was forced to resign.

He then notified VEPCO and the Atomic Energy Commission (AEC) and still received no response. He notified the office of the governor of Virginia and the Virginia Department of Labor, still with no result. Finally, his two senators from Tennessee managed to convince the AEC to investigate his allegations. The AEC confirmed Houston's charges. Finally, VEPCO hired a consulting engineering firm, which also concluded that Houston was correct in finding welding deficiencies. Finally, the AEC required that the plant have three times the inspection frequencies of normal nuclear plants.

Houston was correct but out of a job. He suffered significant financial losses due to lack of employment and legal expenses.

Being Right Is Expensive

In the mid 1960s George B. Geary worked essentially as a sales engineer for U.S. Steel in the tubular steel products area. Though not an engineer by education Geary had developed, through 14 years of working with U.S. Steel, a substantial background in the technology of manufacturing steel pipe. At this time the company was introducing a new product for the oil and gas industry, but Geary thought it had not been adequately tested and might rupture under high pressure. He so informed his superiors and suggested that additional testing be done before selling the pipe, thereby preventing financial loss due to personal injury and damage suits as well as loss of reputation for U.S. Steel. His immediate superiors insisted he proceed to market the pipe without additional testing, and Geary did so. He also went above his superior's head to corporate headquarters. Ultimately, a vice president thought enough of Geary's idea that he ordered a suspension of the sales effort until further tests were run.

On July 13, 1967, George Geary was fired. U.S. Steel even tried to prevent him from receiving unemployment compensation while he was looking for a new job, saying he was discharged for willful misconduct. Nearly a year later, the Pennsylvania Board of Review concluded that he had the welfare of the company in mind and could not be charged with willful misconduct; thus he was entitled to unemployment compensation. He tried suing U.S. Steel for wrongful discharge but lost his case in a closely decided case, split 4 to 3.

It seems that it takes a heroic effort to fight a company decision, even in matters of safety. This certainly could be a moral dilemma some of you will face. It is an unusual situation that will reach the dramatic instances detailed above, but not an impossible one. Engineering societies have proposed that legislation be enacted to prevent firing an engineer when his or her acts are consonant with the ethical obligation of a professional to hold paramount factors relating to public safety, health, and welfare. This legislation is making slow progress in terms of judicial trends and legislative initiatives.

Another more dramatic proposal is that if an engineer's judgment is overridden by a manager, and the engineer formally objects, arguing that this would create a serious danger to human safety or health or perhaps lead to serious financial loss, and if the manager persists in his or her decision, responsibility for the consequences rest with the manager. The manager would be legally liable if in the future such danger or loss occurred. As it is now a manager has no legal responsibility for actions associated with decisions in these areas. Liability would tend to reduce the temptation to make decisions for immediate gain, such as meeting deadlines and cutting costs. Furthermore, it would make managers responsible for their actions whether or not they still worked for the company when the real effect of the decision, such as a plane crash or other disaster, occurred.

Whereas this interesting idea has been proposed by others, the process of drafting such legislation would be formidable, but perhaps society needs initiatives in this area. It again points to the fact that engineers have to become more politically and socially aware as well as being technically competent.

A Positive Side to Perseverance

The previous examples illustrated some of the negative forces you may confront as a practicing engineer, but it is important to note that speaking out as a concerned citizen does not usually result in dismissal actions.

In 1977 James Creswell, an inspector for the Nuclear Regulatory Commission (NRC), was assigned to inspect the start-up conditions at a new nuclear generating station, Toledo Edison's Davis-Besse facility. During the low-power testing a sudden and significant generation of heat occurred due to failure in the main feedwater system. The reactor operators did not receive clear information as to what was happening within the plant, because of faulty instrumentation and control systems. The operator assumed a valve was stuck in the open position and closed down the emergency core cooling pumps.

A similar mishap had occurred at a nuclear generating station in California, though the cause was completely different; misinterpretation of information and insufficiently clear information from the instrumentation and control systems caused operator error that could have resulted in a major catastrophe. Fortunately, none occurred at the Davis-Besse plant, for it was operating at 9-percent power and the actual valve failure was detected after 22 minutes, a relatively short time.

Creswell was disturbed that luck should be a factor in the safe operation of nuclear reactors and for over a year communicated his views to all parties concerned—the NRC, the utility, and the manufacturer of the power plant. No party was interested. He persisted. Finally, he took a day's leave and at his own expense traveled to meet with two receptive NRC commissioners in Maryland. They listened and subsequently requested that NRC staff members answer the questions Creswell raised. As the memo was being typed, the Three Mile Island disaster occurred.

Unlike Davis-Besse, Three Mile Island had been operating at 96-percent power and the failure of the relief valve to close had taken over two hours to detect. This combination of operator errors created an explosive situation, in part because of deficiencies that Creswell had noted. He later received a $4000 award from the NRC. While his concern did not prevent this tragedy, his efforts and concerns for public safety were recognized. In the final analysis, his perseverance paid off.

Holding the Line

Probably everyone is familiar with the explosion of the Challenger space shuttle. The simple fact is the shuttle flew against the advice of engineers, particularly Roger Boisjoly and Arnold Thompson. They stated that the anticipated cold weather conditions in which the shuttle was scheduled to fly could create a situation where the rocket booster seals would fail. That is exactly what happened, causing a fatal explosion.

Higher management in Morton-Thiokol and NASA overruled their advice. Boisjoly and Thompson could not prove that it would happen, only that it might happen, and it was important to Morton-Thiokol and NASA to have the shuttle fly.

Following the disaster, Roger Boisjoly testified before the federal investigative commission. There was a great deal of sentiment against him, and he was able to keep his job in part because of Congressional pressure, though his status within Morton-Thiokol suffered. He offered these words of advice to engineering students at MIT the following year: "I have been asked by some if I would testify again if I knew in advance of the potential consequences to me and my career. My answer is always an immediate yes. I couldn't live with any self-respect if I tailored my actions based upon potential personal consequences."

This addresses a fundamental reason many engineers practice their profession. It contributes directly to their sense of value, their level of happiness. Aristotle noted that happiness is self-realization, not contentment. Having an easy life, according to Aristotle, is not the path to happiness; rather, using your abilities, your talents, skills, and interests, to their fullest yields the path to happiness. Continuing along this line of reasoning, it follows that the more complex and challenging the situation in which you use your abilities, the greater your happiness. Certainly the undergraduate engineering curriculum is one of the most, if not *the* most, challenging and complex academic paths available. There is a great deal of satisfaction in undertaking this challenge in school and in your pursuant career—one reason many intelligent people find engineering very rewarding.

4.5 DESIGN CHANGES

Changes are often made in the execution of a design. As an engineer in charge of a project or product, you must make sure that the changes do not affect the product or project integrity. A case in point where the changes affected the project was in the construction of the elevated walkways in the Hyatt Regency Hotel in Kansas City, Missouri. Two elevated walkways, shown in Figure 4.1, spanned the lobby. In July 1981 a dance contest was in progress and over a thousand people were on the overhead walkways, watching the participants below, with some of the observers dancing as well. The walkways were not constructed as designed but had loads twice that of the original design. The walkway supports failed, and 111 people were killed and 188 injured. Several factors led to the failure of the walkway, compounding the effect of any one factor. They included the dynamic load caused by dancing, a welded seam that had to be load-supporting, and the possible omission of some load-bearing washers.

The combination of all events caused the failure, and the constructed walkway was of a different and poorer design than the original. Thus, not only when changes are made should all safety considerations be viewed, but the product or project as constructed must be inspected to make sure it conforms with the design specifications.

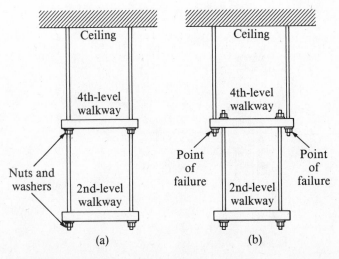

FIGURE 4.1 The walkway of the Hyatt Regency Hotel, (a) as designed and (b) as constructed.

4.6 STATE-OF-THE-ART IS NOT ALWAYS ENOUGH

Disaster may strike even when engineers have done everything correctly. Traditional methods are followed, the plans are constructed as designed, but at some point the product or project fails. Perhaps an aspect of the physical world is exposed for the first time in the new design; the fault does not rest with the engineer but with the body of knowledge that must be extended. Remember, engineers design by making assumptions about how nature behaves; when designs are pushed to their limits, the model of nature may not be sufficiently accurate and must be redefined.

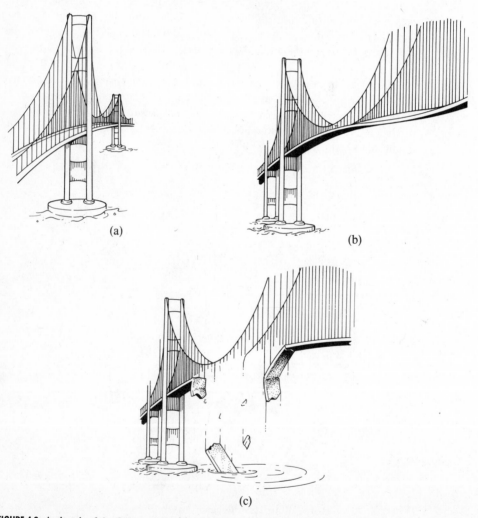

(a)

(b)

(c)

FIGURE 4.2 A sketch of the failure of the Tacoma Narrows Bridge: (a) the initial design, (b) the bridge undulating due to wind and vortex shedding, and (c) the bridge collapse.

(a)

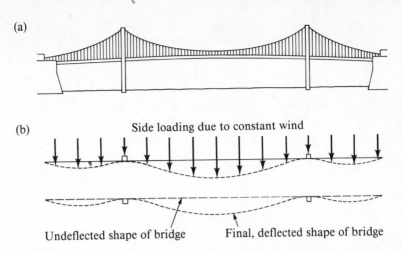

(b)   Side loading due to constant wind

Undeflected shape of bridge   Final, deflected shape of bridge

FIGURE 4.3 The Tacoma Narrows Bridge, (a) side and (b) top views.

A classic case in point is the Tacoma Narrows Bridge, completed on July 1, 1940; it self-destructed four months later. The bridge had a center span of 2800 feet and a width of 39 feet, making it the slenderest bridge ever built, although the fabrication and design techniques were standard. A strong crosswind blows at the narrows, and hit the bridge broadside. This caused the bridge to undulate during strong winds. In winds of only 42 miles per hour, the undulations became so severe that the bridge collapsed. Figures 4.2 and 4.3 illustrate the process. Fortunately, the problem was recognized in advance, so that people were not allowed on the bridge, and no loss of life occurred. Wind tunnel testing showed that vortices formed alternatingly on the upper and lower surfaces of the bridge, causing it to vibrate. At a wind speed of 42 miles per hour, the vibrations coincided with the natural vibrations of the structure, causing it to rip apart.

4.7 SAFETY AND RISK

Society, the collective consumer, would like all products and projects to be 100 percent risk free. That is never possible, but a design should present minimum risk to people and the environment.

We are making personal risk/benefit decisions all the time, deciding to smoke, driving a motorcycle, eating certain foods; these actions are our choice, we determine them. In determining the location of a waste disposal facility, however, an engineer chooses for us, assessing the risk after examining a variety of effects and their probability of occurrence. The engineer's choice of acceptable risk may be reasonable yet completely unfeasible, given the way the public perceives certain risks. As an example, the Dutch people living near the dikes risk one death per ten million per year, or 10^{-7}. Chemical plants with a similar risk ratio were to be built near them, and the people strenuously opposed the plants. The citizens felt that any risk from constructed hazards was unacceptable and could not be reduced to statistics. Similar feelings are associated with nuclear power.

If we look at areas other than hazardous materials, the matter of risk assessment is less psychologically loaded. In developing a product the applicable regulations, codes, and standards are identified, and these establish the basic safety requirements. The product concept is weighed with these on one side and the schedule, feasibility, and costs on the other side. The designer must examine safer alternatives if they exist and if a product liability lawsuit is a possibility. The engineer must consider the user in designing any product or project. This becomes very difficult for a large project, as the complexities of the systems create a staggering array of possibilities for misuse.

Disasters at Three Mile Island, a nuclear power plant; Chernobyl, a nuclear power plant; and Bhopal, a chemical manufacturing plant, were all caused by human error. A combination of events, seemingly impossible and therefore not designed for, as a combination of safety and operating procedures were ignored, produced the disasters.

4.8 SITUATIONS YOU WILL FACE

You will probably face several situations involving ethical standards when you look for a job upon graduation. The first involves the recruiting process.

As you near graduation you send out résumés, and two companies, near each other but a distance from you, have asked you to visit them for a job interview. The companies will reimburse you for your expenses, airfare, hotel, car, and meals. Should you tell the companies that you are interviewing with both of them? Yes, though you need not specify who the other company is. You must divide the costs associated with the visit between the companies, so each pays its fair share. The cost of the interview to you is your time, but to them it's time and money. You should not charge each independently, making income for yourself on the interview process. This is the heart of the matter: not disclosure, but that you not use the visit for personal gain.

Another situation you may have to contend with is multiple job offers, but at different times. Consider the following. You have interviewed with companies A and B, and while the opportunities with both are good, you prefer those of company A. You receive a job offer from company B and have to respond in two weeks. You find the offer acceptable but hope in the two-week period to hear from company A: no such luck. Deciding to take a bird in the hand, you accept B's offer. Three weeks before you are to report, company A makes you the offer you had been hoping for. What do you do?

Accepted practice is that as long as you have not started work for the company, you may notify company B that you can no longer come to work for them and accept company A's offer. This should not be used as a bargaining tool, nor should you be under any financial obligation to company B, such as their having sent you moving expenses or having found housing for you. If you have started work for company B, it is improper to accept the offer from company A and quit work. Once you start working, company B is investing in you, training you. A certain time frame is involved before you return this investment to them, perhaps a year. Should company A still be interested in you after a year and you feel the same, then a switch is possible and ethical.

What about switching employers—how much notification should you give? Enough, but not too much; normally, between two weeks and one month. You want the company to be able to look for a replacement for you. On the other hand, giving too much notice creates an awkward situation with your fellow employees. You are leaving the group and that distances you from them; your focus is where you are going and theirs is where they are. Even though employment changes are expected by the company, staying beyond a month often creates uncomfortable tension in the workplace.

What can you do now to prepare for your future career in engineering? Being an ethical and responsible professional does not start with graduation. It has to be developed now. Honesty and integrity are important attributes everyday, in completing homework assignments, writing papers, taking examinations. Trustworthiness is necessary when taking an examination, as well as when making difficult ethical decisions as an engineer. Many of the moral dilemmas we face are brought about by paying attention to shallow values, such as pay, promotion, status, power, and possessions, instead of paying attention to the more significant values of human dignity, compassion, and self-esteem.

4.9 CASE STUDIES

The following situations are drawn from articles appearing in *Engineering Times,* a publication of the NSPE, for general information to engineers. This material is reprinted with the permission of the National Society of Professional Engineers and *Engineering Times.*

Case I: When Is a Conflict of Interest Not a Conflict of Interest?

(*Author's note:* Before reading the case refer to the NSPE Code of Ethics in Appendix 1, particularly Sections II.4.e and III.2. These deal with conflict of interest and responsibility to society.) Here is the situation.

The law in a certain state requires every community to have a municipal engineer, whose duties and pay are fixed by municipal ordinance. Those duties vary according to the size and type of city but generally consist of attending official meetings, providing advice on engineering matters, maintaining tax maps, reviewing site plans and subdivision maps, preparing cost estimates for proposed facilities, handling complaints from citizens, and giving advice on what consultants might be needed for projects.

The city of Middleville cannot afford a full-time municipal engineer or the support staff needed for a full-time office. Middleville hires A-Z Engineers for a relatively low fee and appoints A. Z. Smith, P.E., a principal in the firm, as the municipal engineer.

Later, Smith's firm submits a proposal on a municipal contract with the city of Middleville.

What Do You Think? Is it ethical for Smith, as municipal engineer, and his firm to submit a proposal on a municipal contract? Smith decided he did not have a conflict of interest from his role as a part-time consultant to the city and submitted a proposal, competing with another engineering firm. Did he have an unfair advantage in making the proposal?

What the Board Said The NSPE Board of Ethical Review said it was ethical for Smith and his firm to seek the contract.

Here's Why The public interest in this case can best be served by providing small municipalities the best engineering services available. And, apparently, the state law is intended to achieve that end.

Continuity of municipal engineering services tends to ensure the best work for the city, assuming that the best available engineers are used. The engineer-consultant who becomes "municipal engineer" should, therefore, not be barred from furnishing the city with complete engineering services through his own organization if he is qualified to do the work.

Case II: Being Paid to Take a Job

(*Author's note:* Refer to Section II.4.c of Appendix 1 before reading this case.) Here is the situation.

Caroline Specs, P.E., a recent engineering graduate seeking employment, receives two job offers: a direct one from RZO, Inc., for a position in its sales department, and the other from Zephyr, Inc., through an employment referral firm, for a position in the design division. Specs finds the type of work at Zephyr more to her liking, but the proposed salary is $5000 a year less than that offered by RZO. After some discussions, the employment firm tells Specs that Zephyr would not increase her proposed starting salary during the first year because of its salary system for all newly hired engineers, but that the employment firm itself would pay Specs an "acceptance bonus" of $5000 if she took Zephyr's offer. She agrees to the arrangement proposed by the referral firm.

What Do You Think? Does this bonus arrangement mean Specs improperly received compensation from the employment agency? Was it an ethical decision?

What the Board Said   The NSPE Board of Ethical Review decided that Specs's acceptance of the referral firm's proposal was ethical.

Here's Why   Although the NSPE Code of Ethics states that engineers should not accept payment from either an employee or an employment agency for providing a job, the apparent intent of this provision is to prevent kickbacks from job hunters to employers. The salary offered should be the full salary without rebate or reduction.

In Specs's case, if the employment referral firm wishes to reduce the amount of its own fee by $5000, it may do so as a business decision, and that decision is not controlled by the code. There is no ethical reason to prevent Specs from choosing her preferred employment and coming out ahead financially by accepting the arrangement.

Case III: Unfair Competition? To Whom?

(*Author's note:* This case is from the *Engineering Times'* "Legal Corner," which gives capsule responses to legal questions members pose. This case has ethical overtones as well. Company X may require you to sign a form that you will not work for competing companies for a certain period if you leave the employ of company X. A major concern is that information you developed or learned at company X can be very useful to another firm, and they might hire you to obtain that information. Read Section II.1.c of Appendix 1.) Here is the situation.

I'm employed by a engineering consulting firm in Florida. Upon entering that job several years ago, I was asked to sign an agreement stating that, in the event I left that firm, I would not compete directly with it. At that time, I didn't fully consider the implications of my signing such an agreement. Now that I'm about to leave the firm, I'm concerned about the restrictions that the agreement might impose upon me. What guidance can be provided?

The enforceability of "covenants not to compete" depends on a variety of factors, the first being whether during the period of employment you, as covenantor, were privy to proprietary information, trade secrets, or other sensitive information of the firm and, if so, whether you plan to use that information in competition with those consultants. It's not always easy to demonstrate the existence of a trade secret, and the firm may have some difficulty in defining what is confidential. If the firm has not taken steps to maintain confidentiality (e.g., it has provided liberal access to the information), your former employer may have difficulty enforcing the covenant.

Another important issue is whether the covenant is reasonable in scope, territory, and duration. For example, a covenant not to compete that prohibits former employees from competing (1) in technical areas that were not the subject of their former employment, (2) in an unreasonably large geographical area (e.g., in all of Texas, California, or the United States), or (3) for an inordinately long period of time, may be found to be unreasonable, and its enforceability may be limited.

Finally, some courts have refused to enforce "covenants not to compete" on public-policy grounds. Those courts generally hold that such provisions lack "mutuality"—employees are giving something up but they are not getting anything in return—and are inconsistent with open competition.

REFERENCES

1. Alger, P. L., Christensen, N. A., and Olmsted, S. P., *Ethical Problems in Engineering,* John Wiley and Sons, New York, 1965.
2. Grassian, Victor, *Moral Reasoning,* Prentice-Hall, Englewood Cliffs, N.J., 1981.
3. Unger, S. H., *Controlling Technology: Ethics and the Responsible Engineer,* Holt, Rinehart and Winston, New York, 1982.
4. McCuen, R. H. and Wallace, J. M., *Social Responsibilities in Engineering and Science,* Prentice-Hall, Englewood Cliffs, N.J., 1987.
5. Nef, J., Vanderkop, J. and Wiseman, H., *Ethics and Technology,* Thompson Educational Publishing, Toronto, 1989.
6. Martin, M. W. and Schinzinger, R., *Ethics in Engineering,* 2nd edition, McGraw-Hill, New York, 1989.
7. Hoover, H. H., *Memoirs of Herbert Hoover, Vol. 1, Years of Adventure,* Macmillan Publishing Company, 1951.

PROBLEMS

4.1 Using the attributes of professionalism, evaluate a physician and a salaried baseball player as professionals.

4.2 Obtain a copy of the Hippocratic oath taken by medical doctors and develop a version for an engineer.

4.3 Often engineers view the term *problem solving* as a process involving a mathematical solution. Propose types of problems in engineering that are not mathematical, but require an understanding of the arts and humanities.

4.4 You are an engineer working for an oil company. The company locates a gas field near a historic trail used by settlers in the westward migration. Local citizens are upset by the prospect of drilling and transporting the gas, as the trail is of historic significance and the ecology of the trail will be destroyed. Discuss the issues involved and factors that you would use in trying to resolve the conflict.

4.5 Referring to the NSPE Code of Ethics, formulate a code of ethics for student conduct in an academic environment.

4.6 The NSPE Code of Ethics suggests there are ethical responsibilities required of the engineer, the employer, the client, society, and the profession. Identify them.

4.7 It is possible for an action to be legal, but unethical. Discuss this using examples.

4.8 An engineering firm has signed a contract for the design of a small flood-control dam. The firm does not have someone on staff with this specific design experience, nor does it have the budget to hire an expert. It assigns the task to someone whose experience is nearest to that needed to design the dam. Discuss the decision in light of the NSPE Code of Ethics.

4.9 Discuss whether the NSPE Code of Ethics implies that engineering firms should provide engineers employed by the company with time off and tuition compensation for advanced degrees.

4.10 You are working for a large company and your supervisor tells you to charge your time to a project other than the one you are working on because its budget has been expended. Does this violate a section of the NSPE Code of Ethics? What are your courses of action?

4.11 Risk assessment is a major factor in value conflicts, as shown by the example of the chemical plant in Holland. Explain the social responsibilities engineers must exercise in developing new technologies.

4.12 List six things you value most. Try to determine how you arrived at this list, what in your background contributed to it—parents, religion, school, self-reflection.

4.13 Describe your personal ethics.

4.14 You have an opportunity to work for a defense contractor at an attractive salary with many of the benefits you want in a professional workplace. A problem with the position is that you will have to work on weapons systems. What ethical questions should you resolve, if any?

4.15 Discuss why a professional person—engineer, attorney, or physician— does not bid competitively on the performance of a service.

4.16 You and a colleague are working on a technical paper to be presented at a national conference. Before the paper is presented, your colleague is fired from the company, and your superior wants you to give no recognition of your colleague's contribution to the paper. Several alternatives are open to you: give the paper as the sole author, give the paper acknowledging your ex-colleague (risking your position with the company), do not present the paper, or take some other action. Discuss what you would do.

4.17 Jane Smith, a conscientious citizen and a principal in a consulting engineering firm, attended a town meeting where a $10 million water pollution control project was under discussion. The town was not certain it could afford the project, but would lose needed federal funds for this and other projects if it did not build the project. Smith proposed to the town that her firm would look for cost savings in the construction plan and would not charge the town anything for her services if she found no savings. The firm would charge 10 percent of the savings if it found any. In no way would it displace the design firm that drew up the plans and specifications. Was Smith's offer ethical?

4.18 A criticism of engineers is that, while they design and create devices, structures, and systems, they are directed to do so by others who, not necessarily engineers, are directing the development of engineers. Discuss the criticism pro and con, including the definition of an engineer and whether a distinction can be made between an engineer and a technician.

4.19 An automotive specialist owns a successful repair shop and has a good knowledge of complex electronic, electrical, and mechanical systems and devices. Discuss why this person is not an engineer.

4.20 Nearing graduation, a college senior decides to interview at a company in another state for a job that he has no intention of taking. He accepts the expense-paid trip. What elements of the NSPE Code of Ethics are involved?

4.21 When one rationalizes, one makes up false motives for one's actions so there are no regrets for those actions. A consulting engineer rationalized her actions for giving a kickback as follows: It was a business decision to make the kickback; no one was injured by my actions. In fact the public received a well-designed project that was completed on time. Plus, without the kickback I would have had to lay off some employees, who would then collect unemployment. Discuss her reasoning.

COMMUNICATION— WRITTEN AND ORAL

■ To explore the functions of communication.

■ To learn how to prepare reports and papers.

■ To write an effective résumé.

■ To understand the preparation necessary for oral presentations.

■ To improve your oral delivery.

■ To use effectively the technical information sources in your university library.

(Photo courtesy of Scott, Foresman.)

Engineers must be able communicators and proficient with oral and written forms of communication.

5.1 INTRODUCTION

The assumption made about most engineers is that they are very analytical, but have poor communication skills. Not only is this assumption invalid, but any successful engineer must be able to present the results of his or her investigation to others. It is important to develop your skills and abilities in this area while in college, as they will be a tremendous asset while you are in school and in your ensuing career.

Engineers must communicate more than many other professionals. For instance, suppose you are working on a project with several people in a large aerospace company. You will report at weekly meetings with your project leader on the progress that is being made. Superiors have to be informed of problems, successes, and the timeliness of the work being completed. This requires communication. It may take the form of an oral presentation at a weekly briefing meeting, or it may take the form of a written progress report. It should be apparent that the better you present yourself through clear presentations and reports, the more likely you are to be promoted. Of course your analytic abilities are extremely important; good communicative ability will show them in the best light.

5.2 FUNCTION OF COMMUNICATION

The purpose of communication is the transmission and reception of information, of ideas and results. Communication does not exist without transmission and reception. The major mistake most people make in this regard is disregarding the receiver of the information. You must make certain that this person is in a position to understand the information as you present it. If the person is a generalist, without detailed knowledge of the analytics, do not focus on the technical aspects, but rather on the ramifications and implications of the results and the effect they will have on the project. To a technical superior you would want to communicate the analytic aspects, assumptions made and reasons for them, in modeling a certain electronic circuit or mechanical system.

The level of the language used is of vital importance. The more obtuse or grandiose it is, reflecting your command of the language and nuances of thought rather than addressing the matter at hand, the greater the likelihood that it will be misunderstood or disregarded. Engineers like straight expository writing. Be direct and to the point so there will be little cause for confusion. Engineers have an advantage over students from other disciplines; in the business world directness, even a certain bluntness, is admired. Not all academic disciplines encourage it, nor is it necessarily appropriate to the education of a philosophy major.

5.3 WRITTEN COMMUNICATION

Precision in report writing is very important, as your company most often will be subcontracting work to others or receiving contracts of work from others. This work must be communicated precisely in job or contract specifications. The specifications will detail exactly what must be accomplished, how much the company will pay or receive, and sometimes the manner in which the work must be done. Interim reports may be required. It should be apparent that being able to communicate clearly and precisely so there is no confusion for either party is vital to operating a profitable business.

Writing is not easy, even when you are skilled at it. Before writing the first draft, think about what you want to present, how to do it, and who the audience is. The draft is edited and finally rewritten. All successful writers, be they novelists or engineers writing technical reports, use these techniques.

As you are preparing to write a paper, a report, or a text, you must think about what it is you wish to communicate. This is the general theme of the paper. Write this down. It requires you to focus your thoughts. If it is a paper, can you in a few sentences tell someone what it is you are attempting? Then move to the specifics. Jot down the ideas you wish to include as part of this general theme or topic. There need not be any order to these; let them flow. Do not break your creative thought processes. They will not form an interconnected pattern at this point, nor is it necessary that they do so. Next organize the specific points in a coherent pattern and think about how they interconnect to support the main topic. At this point you will probably have additional points you wish to make, or amplifications of existing ones. Insert these. At last you can begin to write: you have a focus, the theme or topic, and you have prepared yourself with the key points you wish to use in support of the topic. Do not be overly concerned about spelling, grammar, and paragraph structure at this point. Once a rough draft is written you can correct these defects. As you read the rough draft, correct for spelling and grammar errors. You will need a dictionary—everyone does. Use it. Nothing ruins a well-thought-out paper faster than misspellings. All writers have a dictionary at hand. A very important feature of word processing software is the "spell checker," which alerts the user to misspellings. Once you have edited your rough draft and inserted any corrections or additional comments, write your final copy. Word processing on a personal computer has a tremendous advantage because of the ease of inserting corrections without having to retype the entire paper.

5.4 THE WRITTEN REPORT

Throughout your professional career as an engineer, including your undergraduate education, you will be called upon to write reports. Most English composition classes do not cover this. A report is a researched document that relies on the incorporation of opinions of others as well as your own. Frequently, the reports you will be writing in college describe laboratory situations where experimental data must be presented, as well as the rhetoric sections.

All written reports should include certain elements:

   I. The front matter
       a) title page, followed by a blank page
       b) abstract
       c) table of contents
       d) list of tables
       e) list of illustrations
  II. The text
       a) introduction
       b) main body of report
           procedure, results, discussion, conclusion, recommendation
 III. The reference matter
       a) appendices
       b) bibliography

Not all reports will contain all these elements; however, a formal report will. Often sections of the main body are combined, such as results and discussion, or conclusion and recommendation. Since the abstract often serves as a summary, an additional summary section is not necessary. In very long reports that are intended for a general and disparate audience, an executive summary of several pages may be created, giving an overview of the report, noting salient points. However, you will not encounter reports of this nature immediately in your career.

Any report you prepare should be typed. If you do not know how to type, learn. Your grade will be a function of how good your report looks to the reader, as well as how cogently it is written. This may seem unfair, but check your own reaction to a paragraph that is handwritten on lined paper or typewritten. Which would you rather read? If you have a personal computer, acquiring word processing software is well worth the expense and will improve the quality of the written reports and papers you must write during your college career. Software packages are available to teach you to touch-type, a valuable and time-saving skill.

Let's analyze what should go into a typical report that will have the aforementioned elements.

Title Page

The title page includes the report title, the authors' names, the date, and the organization's name and address. For student reports the organization's name is the department and college. The organization in general will be your employer; the date is important so the reader will know the timeliness of the information in the report. The information should be correctly capitalized, centered, and spaced attractively on the page.

Abstract

Although the abstract is one of the first elements of the report, it is one of the last elements that is written, and often one of the most demanding. The abstract should be no more than two hundred words, often half that amount. It must contain a description of the problem being analyzed, the method of analysis, and the principle results arising from the study. Quite frequently the abstracts of technical papers and reports appear in other publications used as reference and research sources. Here is a representative abstract.

> This paper studies the rapid shearing flow of dry metal powders. To perform this study, we built and used an annular shear cell test apparatus. In this apparatus the dry metal powders are rapidly sheared by rotating one of the shear surfaces while the other shear surface remains fixed. The shear stress and normal stress on the stationary surface were measured as a function of three parameters: the shear-cell gap thickness, the shear-rate and the fractional solids content. Stresses are measured while holding both the fractional-solids content and the gap thickness at prescribed values. The results show the dependence of the normal stress and the shear stress on the shear-rate. Likewise, a significant stress dependence on both the fractional solids content and the shear-cell gap thickness was observed. Our experimental results are compared with the results of other experimental studies. (*Journal of Applied Mechanics*, 53:935)

Notice that the abstract has all the necessary elements: the problem of rapid shearing of powders, the method of solution with the experimental test facility, and the principal results concerning the normal and shear stresses.

Table of Contents

As in this text, a report should have the major headings and subheadings listed in the table of contents with the appropriate page number. This allows access to those sections of immediate importance to the reader.

Lists of Tables and Illustrations

The headings or titles of tables and illustrations should be given, with the appropriate page number. When there are many tables and illustrations, they are not integrated with the text but separated at the end of the report.

Introduction

The introduction is the first page of the formal report and the first topic. It should briefly acquaint the reader with the problem analyzed, what the history of the problem is, how others have tried to analyze the problem, and what you are attempting.

Procedure

The procedure section is perhaps the dullest part of any report, but critically important in that the assumptions inherent in your model used in the analysis are stated and justified. If the report concerns experimental work, you will detail how the experiment was devised and constructed and the procedure for gathering data. Computer programs will be placed in the appendix, but the methodology in developing them will be covered here. This section is essential for the reader in assessing the validity of the assumptions—you are demonstrating your technical prowess here.

Results

The results section yields that critically important piece of information—what your analysis, experiments, or survey yielded. These next section discusses these results, which is why the two are often combined in smaller reports.

Discussion

In the discussion section you give your analysis of the results. Are they as you expected them to be in light of your analysis? Why or why not? A discussion of the error involved in the results should be presented here. Remember there is an error associated with any measurement, with many computer programs, and with statistical analyses used in sampling data. You should discuss the magnitude of the errors and the effect they have on the results. In Section 7.6 we will discuss error analysis and why a scatter of data is to be expected.

Conclusions

In the results and discussion sections you made no assessment as to meaning of the results. Those sections gave the facts and an analysis of their validity. The conclusion is the most difficult section to write for many students and engineers because here you must apply your engineering know-how to explain why the results are as they are. You will be combining analytical, or theoretical, considerations with the actual results to reach a meaningful judgment. In the situations you will most immediately face, laboratory reports, you will be expected to synthesize the theoretical expectations with the experimental data to explain the results. For instance, in a physics laboratory you are experimentally determining gravitational acceleration. The number you calculate from your experimental results will be different than the value given in the text. Why? In this case perhaps error analysis provides the answer. What is your conclusion regarding the experimental technique? Can your value of the acceleration be of any use?

Recommendations   In most laboratory reports recommendations are not expected, but in engineering reports this section can be the most important one to the company. Should they go ahead and manufacture a new product, will its quality level be sufficiently high, will it function well over its expected lifetime? The recommendation should reference the results and conclusions to show a logical connection and development.

Appendix   The appendix and bibliography or reference section follow the tables and illustrations; either the bibliography or appendix may appear first. The appendix includes material that is too cumbersome to have in the body of the report. For instance, a set of sample calculations would be included here. There is no hard and fast rule about which material should be in the appendix, except that the report should read smoothly and the reader should not be deterred by lengthy calculations, theoretical developments, or data reduction that must be included, but whose value is in its result. This information is suitable for the appendix. For laboratory reports the appendixes would contain your calculations, the data collected, or a computer program listing. For those people very interested in your report, all the information necessary to perform the same analysis should be available in the report.

Bibliography   Your bibliography lists the references you used in performing the report analysis. These will be classical text references, handbooks, and journal articles.

  The following are examples of how to reference a text or handbook and a journal article.

  Burghardt, M. David. *Engineering Thermodynamics with Applications*. 3rd ed. Harper and Row, New York, 1986.

  Craig, K.; Buckholz, R.; and Domoto, G. "An Experimental Study of the Rapid Flow of Dry Cohesionless Metal Powders." *ASME Journal of Applied Mechanics* (1986) 53:935–42.

5.5 RÉSUMÉ OR VITA

A résumé or vita is a brief record of who you are; it will contain your biographical information with particular emphasis on education and work experience. Figure 5.1 illustrates a typical one-page résumé for a senior engineering student. It may seem unusual to include résumé preparation in a freshman course. While it is true your graduation is several years in the future, your career in engineering starts now. You can plan your résumé so that in four years' time it is at least as impressive as the one in Figure 5.1. People receiving the résumé want to be able to connect readily the job requirements with your experience. Each of us is unique, and you will have had experiences that no one else has. Your résumé should portray you as accurately and as positively as

Jane A. Doe

Address: 194 Third Avenue
New York, NY 10022
212/692-1113

Objective: Full-time employment in robotic system development.

Education: Top Name University
B.S. in Mechanical Engineering expected May 19XX.
GPA 3.2/4.0

Honors and   Dean's list: 4 terms
Awards: Tau Beta Pi, National Engineering Honor Soc.

Experience: Top Name University, Advisement Office: September 19XX–present; tutor in sophomore engineering courses for international students.

Marshall Engineering, Inc.: Summer 19XX; assisted engineers in laboratory testing of pneumatic and electronic control system components; wrote preliminary sections of final report.

Fastfood Chain, Inc.: Summers 19XX and 19XX; worked in a variety of positions; promoted to assistant shift supervisor, responsible for work of 8 others.

Activities: Vice President—TN University Chapter of Society of Women Engineers, 19XX–XX
Member, American Society of Mechanical Engineers
Senator, TN University Student Senate
Intramural sports: volleyball, handball
Hobbies: hiking, photography, tennis

Personal: Citizenship: United States of America

References: Dr. Richard Jones
Mechanical Engineering Department
Top Name University
New York, NY 10001

Dr. Sandra Degas
Mechanical Engineering Department
Top Name University
New York, NY 10001

Mr. James Earl
Senior Engineer
Marshall Engineering, Inc.
1025 East Road
Flemington, NJ 15231

FIGURE 5.1 An example of a résumé for an engineering student.

194 Third Avenue
New York, NY 10022
15 April 19XX

Mr. Charles Foster
Director of Personnel
Precision Engineering Co., Inc.
Cambridge, Massachusetts 12062

Dear Mr. Foster:

I am very much interested in working for Precision Engineering Company in the area of robotic systems and have enclosed a résumé of my related work experience.

I would like to amplify briefly certain sections of the résumé to give you a clearer picture of my abilities. Through my work as a tutor, I have gained a true understanding of engineering fundamentals and have found that not only was the opportunity to help others rewarding, but my comprehension of complex subject matter is more complete.

While employed by Marshall Engineering I realized that a career in engineering was indeed the right choice for me. I enjoyed the commercial realities that engineers must confront and worked well with the senior engineers. I feel very proud that they trusted me enough to allow me to prepare the introduction and test sections of the evaluation reports on the G26 and E28 control systems.

I believe that my background shows that I contribute to every organization I am affiliated with, be it student government where I learned the give and take necessary for group decision making, or Fastfood Chain, where I was promoted to assistant shift supervisor at the age of 19. I know that I can contribute to your company and look forward to hearing from you.

Sincerely,

Jane A. Doe

FIGURE 5.2 An example of a covering letter for an engineering student's résumé.

possible. Do not exaggerate, but you can discuss the qualities that show your initiative, responsibility, and engineering skills. You may want to prepare several résumés emphasizing different abilities depending on what the job you are applying for requires.

Let's look at Figure 5.1. Much of the information is standard, and personnel directors look for it. If your grade point average is low, it may be best not to mention it at all. You will have to explain this in your covering letter. Your transcript will have the GPA on it, so the issue must be addressed, but not necessarily immediately. You will have to have other attributes to compensate for the low GPA. In tailoring your résumé, you may wish to change the objective slightly to better meet the expectations of a given company, if you are aware of what they want and it fits with your background.

The work experience noted in the résumé is not extraordinary; virtually any engineering student could have this experience if he or she wished. Not everyone can work for three summers in engineering firms. Many of you will work for stores and other commercial enterprises. Use them to your advantage: if you were promoted, bring it to everyone's attention. This means someone else thought well enough of you to promote you—you are a valuable person. The other employment situations show that you work well with others and have exercised initiative and responsibility.

You will be writing a letter to accompany the résumé. It provides you with the opportunity to briefly amplify certain aspects of your educational and work experiences. Figure 5.2 shows a cover letter for the résumé in Figure 5.1. Remember the purpose of the résumé and the cover letter is not to get you a job; you will have to do that in person. It is to interest your prospective employer in you, to show that you will be the type of employee that his or her company is looking for. During the interview be prepared to expand more upon your résumé. The interviewer will use this as a basis for asking questions and developing a conversation with you to assess who you are as a person.

5.6 ORAL COMMUNICATION

You will frequently need to make oral presentations to colleagues at weekly meetings or conferences. It is quite normal at first to be a little anxious, and you can take steps now that will help you throughout your career. One of the first is

Engineers seldom work in isolation and need to be able to clearly communicate with their colleagues. (Courtesy of Michelin Tire Corporation)

to take a course in speech, perhaps as part of your humanities requirements. In some engineering classes you may have reports to present orally; be sure to participate. Few of us relish the opportunity to do this, but the experience builds poise and confidence. If you cannot give presentations in these areas, look for less threatening areas for initial participation, such as in church or civic organizations. For most of us the nervousness disappears after the first minute or two, especially if we are well prepared and we become involved with the topic under discussion. Few of us can stand up and talk about a subject without preparing what it is we want to say. Just as with writing, you must have a clear idea of what it is you want to express. Initially jot down the facts and the integrating theme that connects them, and build your presentation, oral or written, with this in mind.

Normally you will know who your audience is and what level of presentation is required, and your presentation should be prepared to suit their background. You should not give too elementary a talk, nor a too sophisticated one, as in either case you will lose contact with the audience. Remember communication is a two-way street—transmission and reception.

In most circumstances you will have 15 to 20 minutes for your presentation; too much longer and the audience loses interest. It should be apparent from your presentation that you are secure and in control of the subject matter. You will be asked questions, and this is often the most informative and important part of a presentation. You must be prepared to expand on given topics.

Preparation

How should you give the talk? Use notes, do not read. Beforehand, prepare an outline of the key topics to be covered. Expand on these topics in your presentation. You should have gone over what you want to say many times before actually making the presentation. Once you believe you have all the information needed to make the presentation, prepare note cards from the outline and give a trial talk to yourself. Time it. Remember to speak clearly and not too fast. Once you are satisfied that all the information you wish to convey is in the material you have prepared, try the presentation out on a friend or co-worker. As you are preparing the talk, do not use jokes to make points or to increase the receptivity of the audience. In general they will fail and should not even be attempted.

Visual Aids

Visual aids are a terrific help; they provide something for the audience to read while listening to you. They help you maintain the awareness of the point under discussion. The three methods used in most circumstances are paper charts, often flip charts that are about two feet by three feet; 8½-by-11-inch transparencies; and 35-mm slides. Imagine yourself in the audience: which visual aid is required? Flip charts work well in small groups, 35-mm slides for large groups; for other groups transparencies are a good choice. If you are using transparencies or slides, be sure to view them before the presentation, to be sure that they look the way you want them, that lines are not blurred, for instance. If possible the projection distance for the preview should be the same as for the event.

A second great advantage to using visual aids is that they are your outline and notes. You can expand upon the topics listed on the chart or graph, and often do not have to use notes at all. When constructing the diagrams and charts, keep them simple. They should augment your presentation, but complicated diagrams are sure to turn off your audience. The slide should contain key words, graphs, diagrams, or photographs that enhance what you have to say. Do not read the slides to the audience: if you have nothing to add to what is on the slide, they do not need you—they can read.

Presentation Techniques

Just as the appearance of your written report affects how people view what is within it, so your appearance affects audience reaction to what you have to say. Not only should you look presentable, neat, and well dressed, but you should behave properly. Speak clearly and loudly enough so the audience can hear you; people do not want to strain to understand what you are saying. And speak slowly enough so they can follow. You know what you want to say and what the connections are, but they do not, and you must allow time for these connections to occur.

Talking quickly and softly are signs of nervousness you can be aware of and correct. Avoid the "uhs," the "likes," and the "you knows." Often you will have a podium to hold your notes and to hang on to initially. Stand erect and do not put your hands in your pockets. Some men will thrust a hand in a front pants pocket throughout their presentation, often jingling change. This is a terrible habit, and totally distracts the audience from what is being said.

Important to the audience and to you is maintaining eye contact throughout your talk. Look at people directly, not the same person constantly, but shift your attention from one to another. This keeps you involved with the audience, makes you aware that your message is getting through, and lets the audience know that you are engaging some of its members constantly.

5.7 LIBRARY USAGE

In your career as a student and as a professional, being able to access information about a particular problem is very important. The library is your best source of information in all areas. Being able to use it effectively will enhance your abilities. You will not need to spend undue time trying to locate a piece of information, and you will be able to locate the information that will help solve your problem. There are no courses available to you in library usage, though often the library staff will show you the resources available. Using the library resources wisely will help you immediately in writing term papers, laboratory reports, and senior papers. Following graduation you will have an ability that few of your colleagues will possess. Locating information necessary for a project, rather than reinventing it, is a valuable skill. This skill also prevents failure altogether, if the information is too difficult to reinvent.

An engineer usually relies on one, perhaps two, indexes and a few major journals that specialize in her or his area of interest. Journals are compilations of technical papers that professionals have written in a technical field, such as heat transfer. Various organizations, including professional societies, publish these journals to help disseminate the technical progress that is being made. An index is a periodic publication, monthly or quarterly, that lists by topic the articles that have appeared recently in a variety of journals.

You may be surprised to learn that when you are looking for information on a given topic in research journals it will not be located in one information source. Instead, approximately one-third of the articles of interest appear in primary journals, one-third in related journals, and one-third in unrelated journals. For example, one may find articles on heat transfer in the *Journal of Heat Transfer* (primary journal), the *Journal of Applied Mathematics* (related journal), and the *Journal of Food Science* (unrelated journal). All of these journals could have articles related to the same problem, and without proper searching, repetition of work and delays may result.

Computer-based searches are generally not available to the student or to engineers in smaller engineering firms. Furthermore, you need to define your problem well before using them, and the use of library is your prime means to do this.

It should be obvious that the usual fuzzy attitude toward library usage must be overcome. When information on every subject is increasing at exponential rates, the engineer of today and tomorrow must have up-to-date information acquisition procedures. The engineer who misses articles of concern to his or her work effectively becomes obsolete without being aware of it.

Scientific Encyclopedias and Technical Handbooks

Let us assume that you want to investigate an area that is new to you. The first place to look would be in an encyclopedia such as the McGraw-Hill *Encyclopedia of Science and Technology;* there are others relating to chemical engineering, biology, and so on. The encyclopedia has an overview of the particular subject by an authority in the field and includes references. In conjunction with the encyclopedia, the various handbooks in a particular field may be used (e.g., *Mechanical Engineers' Handbook, Electrical Engineers' Handbook*). They contain a wealth of information in compact form. Often references are cited that are recognized as basic source material. Using these references is the next step; they help you define and better understand the problem before proceeding to current literature.

At this juncture two paths are open. You can consult the card catalog under the particular subject heading or start a search of the current literature.

Card Catalog

Card catalogs will have subject headings for books in various fields, such as diesel engines, thermodynamics, and electrical circuits. Included on the subject card is a brief content summary of the book. Figure 5.3 shows a typical card catalog listing. The title, author, publisher, copyright date (1986), and total

BURGHARDT, M. DAVID
  Engineering Thermodynamics with Applications. 3d
  ed. New York; Harper & Row, © 1986, 608 p., ill.; 24
  cm (07945539). The Harper & Row series in mechanical engineering.
  Includes index.
  Bibliography: p. 516
  1. Thermodynamics I. Title II. Series
CALL NO.: TJ 265.B87 1986 copy 1

FIGURE 5.3 A card catalog listing.

number of pages are listed. The text is part of a series published in mechanical
engineering by Harper and Row. Most importantly, the call number tells you
where to locate the book.

Indexes

To find current works, you will use the various indexes at your disposal. Perhaps
the most familiar to most scientists and engineers is the *Index of Applied Science
and Technology*. This monthly index has various subject headings under which
articles from 335 journals are classified. This index has no author listing; hence,
there is no ready method of checking whether a particular author has published
anything or not. Figure 5.4 shows a typical listing from the *Index of Applied
Science and Technology*. The article title is the key to indexing the article, and
you may be misled or miss relevant articles because of their title; this highlights
the importance of a title when writing a report or paper.

The *Engineering Index* is a second popular index, which selects articles of
engineering significance from approximately 2700 journals and periodic
publications. It includes an author index, so that if you want to know whether an
author who publishes in a certain field has published anything recently, you can
check directly. A further asset of this index is that a brief abstract of the journal
article is given along with the title. Figure 5.5 shows an entry from the
Engineering Index for the same article as in Figure 5.4. Notice the increase in
information that you have available to you. The author's abstract has been edited
in this case, but you have a clearer idea of what is presented. Furthermore, you

Heading: **Laser beam welding**

                  Modelling the fluid flow in laser beam welding. M. Davis
                  and others. bibl il diags *Weld J* 65:167s-74s Jl '86

Explanation: An article on subject of laser beam welding entitled,
                  "Modelling the fluid flow in laser beam welding," by M.
                  Davis and others, with a bibliography, illustrations and
                  diagrams, will be found in *Welding Journal*, volume 65,
                  pages 167–74 in the July 1986 issue.

FIGURE 5.4 A sample entry from the *Index of Applied Science and Technology*.

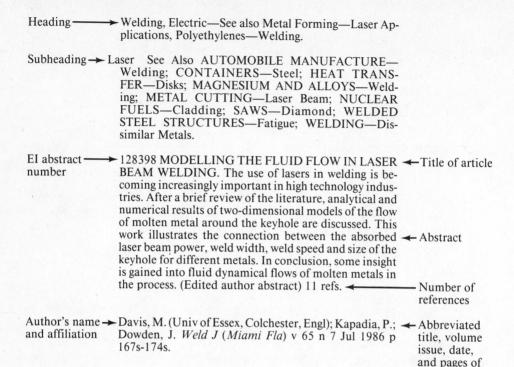

Heading ————→ Welding, Electric—See also Metal Forming—Laser Applications, Polyethylenes—Welding.

Subheading →Laser   See Also AUTOMOBILE MANUFACTURE—Welding; CONTAINERS—Steel; HEAT TRANSFER—Disks; MAGNESIUM AND ALLOYS—Welding; METAL CUTTING—Laser Beam; NUCLEAR FUELS—Cladding; SAWS—Diamond; WELDED STEEL STRUCTURES—Fatigue; WELDING—Dissimilar Metals.

EI abstract ————→ 128398 MODELLING THE FLUID FLOW IN LASER ←—Title of article
number   BEAM WELDING. The use of lasers in welding is becoming increasingly important in high technology industries. After a brief review of the literature, analytical and numerical results of two-dimensional models of the flow of molten metal around the keyhole are discussed. This work illustrates the connection between the absorbed ←— Abstract laser beam power, weld width, weld speed and size of the keyhole for different metals. In conclusion, some insight is gained into fluid dynamical flows of molten metals in the process. (Edited author abstract) 11 refs. ←————— Number of references

Author's name →Davis, M. (Univ of Essex, Colchester, Engl); Kapadia, P.; ←— Abbreviated
and affiliation   Dowden, J. *Weld J* (*Miami Fla*) v 65 n 7 Jul 1986 p   title, volume
                  167s-174s.   issue, date,
                                                                          and pages of
                                                                          source
                                                                          publication

FIGURE 5.5 A sample entry from the *Engineering Index*.

have information to allow you to contact the author if you are pursuing research in a similar area. A limitation to both indexes is a delay of up to a year before the journal article is indexed. It takes time to receive the journals, decide on their heading and the location in the index, input the information, print the index, and distribute the index to the libraries. This is a continual process, and six months' lag time is common.

Delay is inherent in any index, but the *Science Citation Index* overcomes that to a great extent. This index covers approximately 3800 journals and periodicals. The index may be broken into several distinct parts: citations, sources, and subjects. The citation section lists prior articles, by author, that have been cited in current periodicals. It includes the current articles citing the prior one and the total number of citations for a particular article. This gives a key to the relative merit of the cited article. An often-cited article will probably be pertinent and authoritative. Einstein's work is still being applied and thus is often cited. The source section is a listing by author of current articles that have been included in the *Science Citation Index*. The subject section is unique. The key words in the article are permuted. Thus, a journal article such as "Stresses in Submarine Superstructures" would have under the heading "Stress" the key words, *submarine* and *superstructures*. After the key word, the author's name is

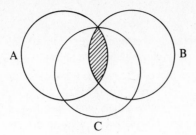

FIGURE 5.6 A Venn diagram illustrating on-line searching of a data base.

given and can be looked up in the source section to find the journal article. The words *submarine* and *superstructures* are also presented as subject headings, with the remaining key words in the title listed below it. The *Science Citation Index* takes some practice to use but is more timely because minimal editing is required before the information is published and distributed.

Government Documents

In this day and age of government-sponsored research, be it funding of a private company or of directly supported research laboratories, no search could be complete without an investigation of the many government reports. Fortunately, the *Government Reports Index,* which indexes all the technical reports monthly, is available. The several different types of indexes include subject, author, and report number. The use of government reports should not be underestimated, as much of basic and applied research is carried out with government funds.

On-line Search

On-line searching of information is provided by a number of services with data bases that have information similar to that contained in the indexes but that have greater searching abilities. These on-line services have two major advantages: they are faster than searching printed documents, and they are more current, as the publishing cycle is shorter—the information is available once it is put into the computer data base. The Venn diagram in Figure 5.6 shows how these services operate. Let's look again at "Stresses in Submarine Superstructures." Each circle, or set of information, will contain the key word *stress, submarine,* or *superstructures.* The intersection of all three sets will yield those articles that have information about all three key words. A computer can determine this quickly and print out the article information for you. For most students the on-line services are not readily available, and you will be using published indexes as your current reference source.

REFERENCES

1. *The Chicago Manual of Style.* 13th ed. University of Chicago Press, Chicago, 1982.
2. Strunk, W., Jr., and White, E. B. *The Elements of Style.* 3d ed. Macmillan, New York, 1979.
3. Turabian, K. L. *A Manual for Writers of Term Papers, Theses, and Dissertations.* University of Chicago Press, Chicago, 1967.

PROBLEMS

5.1 Prepare your own résumé using Figure 5.1 as a guide. Think about what you can do during your college career to enhance your qualifications.

5.2 Look at a simple piece of equipment with no moving parts, such as a window frame, a curtain rod, a door. Write a specification so another person could construct this device. Share your specification with a classmate and see if he or she understands exactly what to do.

5.3 Go to the library and find the encyclopedias, handbooks, card catalog, and indexes mentioned in the text. Locate engineering texts on electrical circuits, electronics, thermodynamics, strength of materials, and aerospace engineering.

5.4 Consider the following general topic areas: robotic arm design, biomechanics and prosthetic devices, advanced composite materials for aerospace application, rocket propulsion systems, integrated circuit design, microvolt measurement in humans, fiber optic transmission, bridge design, and automative engine design. By consulting encyclopedias and handbooks, narrow the area to coincide with your interest, and write a one-paragraph design specification of this area. Then consult the card catalog and list any texts, not more than three, that pertain. List five current references from the scientific and engineering indexes available to you.

5.5 Prepare a five-minute talk on one of the areas mentioned in problem 5.4. Develop flip charts and notes, and listen to yourself give a mock presentation. If possible do this with several classmates, and critique each other.

5.6 Select a journal article that you understand, and write an abstract for it. Does it correspond to the one prepared by the article's author? Why or why not?

5.7 Critique your essay from problem 1.1 in light of the information in this chapter. Rewrite it.

6

GRAPHICAL COMMUNICATION

- To learn the elements of graphical communication that complement verbal communication skills.

- To become familiar with presentation techniques for qualitative and quantitative data.

- To practice the fundamentals of plotting on semilog and log-log paper.

- To use sketching fundamentals.

(Art courtesy of Biblioteca Ambrosiana, Milano, Italy.)

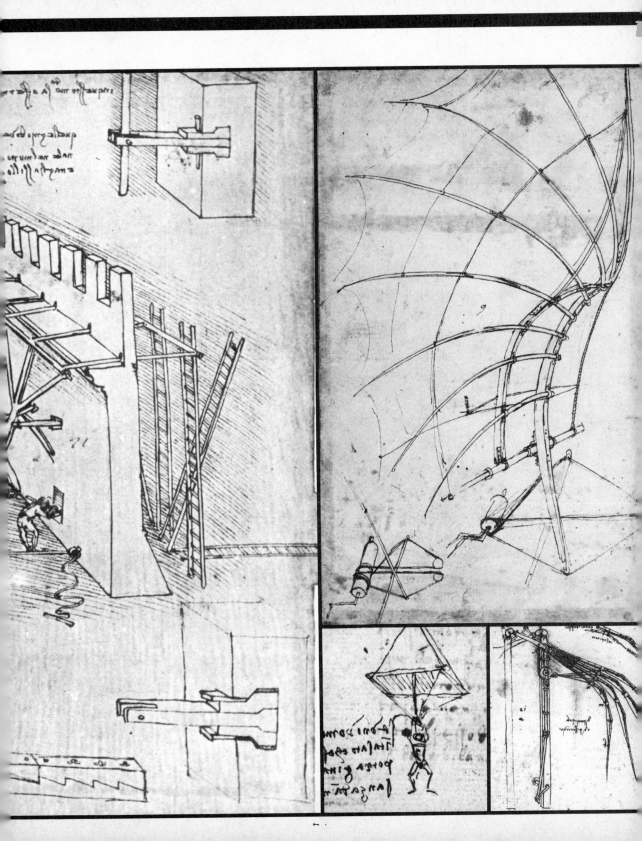

In this chapter you will complement your communication skills with the introduction of elements of graphical communication.

6.1 INTRODUCTION

The adage "a picture is worth a thousand words" is true in engineering. Often a graph can be used to display a trend or a relationship between variables that is not apparent from looking at numerical data. Hence, graphical analysis is an important engineering skill, augmenting written and oral communication skills. Additionally, you will have to make sketches as part of your everyday engineering life. These sketches help you understand a problem you are trying to solve, such as the forces on a certain moving part, and communicate ideas to others. Very often you will want to sketch an object to complement your verbal description. As an engineer you will be responsible for reading and approving drawings made by a skilled draftsman, and although you do not need the skill to produce such drawings, you must be able to read them. These drawings are still called blueprints, referring to the printmaking process in which the lines were white on a blue background. Now, however, the lines are usually blue and the background white. Figure 6.1 illustrates a blueprint of a shipboard refrigeration system. The output plotters used in computerized drafting allow a variety of line colors, but the term *blueprint* will still mean a line drawing, regardless of the color.

In graphical communication you must consider the type of information that you are presenting as well as your audience. We can divide graphical analysis into two general types, one where the information is qualitative in nature, using such techniques as pictographs, bar graphs, and pie charts, and the other where the information is quantitative, using line graphs.

6.2 PRESENTATION OF QUALITATIVE RESULTS

In making a presentation you may want to give people a general understanding of the magnitude of the terms, without presenting the specific numerical data. In such a case pictographs are useful.

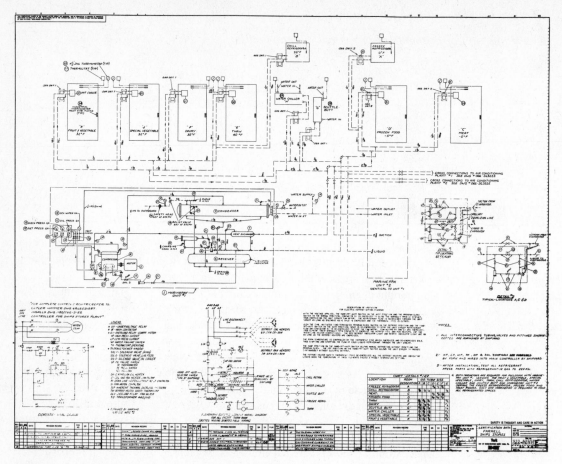

FIGURE 6.1 A blueprint of a shipboard refrigeration system.

E X A M P L E

6.1

The following data represent the world natural gas reserves in trillions of cubic feet. Create a pictograph illustrating the data.

Africa	209
Asia Pacific	120
Europe	142
Middle East	513
North America	288
USSR	918
Other	114
Total	2304

PROCEDURE:

1. Determine the size of the fundamental unit for each picture and what to use as the picture.
2. Determine the number of units required for each item.
3. Construct the pictograph.
4. Label and title the graph.

SOLUTION:

You need to have a picture such that the smallest quantity has some part of the shape and the largest quantity does not have too many items. In this case the largest value is 918 and the smallest is 114, so let 114 be the quantity of the smallest picture. The picture in this case is an oil and gas derrick. When we divide the total production by 114, the following figures are obtained. Note that it is impossible to draw small fractional parts, and they have to be rounded off. This is in part what makes this presentation technique qualitative.

Africa	$209/114 = 1.83$
Asia Pacific	$120/114 = 1.05$
Europe	$142/114 = 1.25$
Middle East	$513/114 = 4.50$
North America	$288/114 = 2.53$
USSR	$918/114 = 8.05$
Other	$114/114 = 1.00$

Figure 6.2 is the pictograph of this data.

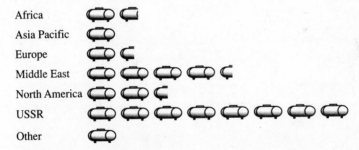

FIGURE 6.2 A pictograph of natural gas reserves by geographic region. Each symbol represents 114 trillion cubic feet of natural gas.

A circle or pie chart illustrates the fraction or percentage of the whole for each component. The circle chart illustrates the size of each component relative to another as well as to the entirety.

E X A M P L E

6.2

Construct a circle chart from the data in Example 6.1, where the circle represents the total world natural gas reserves.

SOLUTION: Determine the fraction or percentage of the whole for each component and then convert this into degrees. The fractions should add up to one and the total degrees to 360. It may be necessary to adjust slightly one or two values so the total is correct.

Africa	209/2304 = 0.0907	0.0907 × 360 = 33
Asia Pacific	120/2304 = 0.0521	0.0521 × 360 = 19
Europe	142/2304 = 0.0616	0.0616 × 360 = 22
Middle East	513/2304 = 0.2227	0.2227 × 360 = 80
North America	288/2304 = 0.1250	0.1250 × 360 = 45
USSR	918/2304 = 0.3984	0.3984 × 360 = 143
Other	114/2304 = 0.0495	0.0495 × 360 = 18
Total	= 1.0000	= 360

Figure 6.3 is the circle chart drawn and labeled with this information.

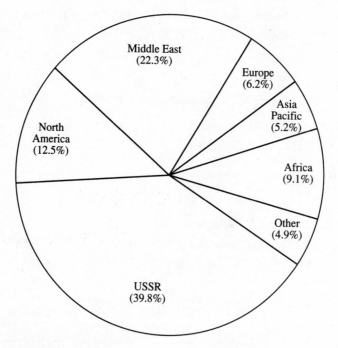

FIGURE 6.3 A circle chart of the world natural gas reserves by geographic region.

The last type of qualitative chart we will address is the bar chart. This chart can be used in the same manner as the pictograph, or can be used to show the variation of quantity over periods of time.

E X A M P L E
6.3

Construct a bar chart to depict the average monthly rainfall in inches from the following data.

Month	J	F	M	A	M	J	J	A	S	O	N	D
	2.5	1.9	2.9	3.1	3.4	2.0	3.9	4.6	3.9	2.7	2.1	2.8

SOLUTION:

Decide whether vertical or horizontal bars will look better. Determine the bar length that allows the maximum amount to be nearly full scale. Next determine the total number of bars, and locate them to cover the space available. Finally, construct the bars, and label and title the chart. Figure 6.4 illustrates the vertical bar chart for the rainfall data.

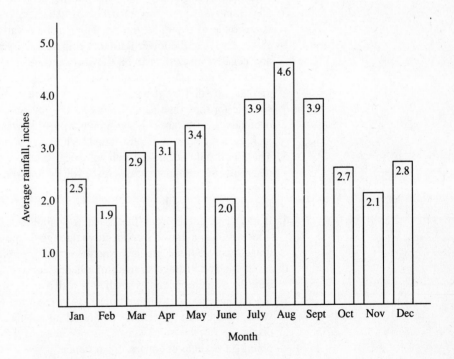

FIGURE 6.4 A vertical bar chart of average monthly rainfall.

6.3 PRESENTATION OF QUANTITATIVE RESULTS

Line Graphs

Line graphs are the most commonly used form for displaying graphical information; you have probably already constructed line graphs as part of some science project. Here we will note some of the rules implicit in constructing the graph and will examine logarithmic graphs, which are often used to present nonlinear data. Graphs are prepared on lined graph paper, normally in pencil, though for permanent work ink is used.

1. Determine if the graph paper should be linear or depict a special kind of function (discussed later).
2. Arrange the data in tabular form for ease in plotting, and determine which variable is the dependent variable and which the independent variable. The dependent variable is normally plotted on the ordinate, with the independent on the abscissa. The chart title will refer to the dependent variable in terms of the independent variable.
3. Determine the approximate locations of the axes, keeping in mind that it is desirable to show the zero location for both variables. Normally the zero value is in the lower left-hand corner of the graph, unless positive and negative values are to be shown.
4. Determine the scale for each axis. Scale equals the range of the variable or data, divided by the scale length available. This number must be suitable for the graph paper available. Do not use awkward fractions or decimals. For instance, if the minimum grid size is 10 divisions to the inch, then the scale divisions should be 1, 2, 5, 10, or a multiple of these.
5. The axes should be marked with numerical values so the reader can easily determine the minimum scale. Also, you do not want the numbers on the axes to be overcrowded; again, this decreases the communicative function of the graph.
6. If the data come from theoretical or empirical equations, plot the information and draw a smooth curve through the points, so the points no longer appear. If the data are experimentally determined, the graph must show this and distinguish this from theoretical or empirical data. In this case the data are plotted as small dots with a circle around each one. The line joining the points should not pass through the circles, so the exact value of the data is clearly visible. If more than one set of data is displayed on the same chart, use different symbols to enclose the point—a triangle or square, for instance.
7. Both axes should be labeled with the appropriate variable description and the units of the variable, for instance, "Velocity, ft/sec," or "Temperature, °K." The axis label should be outside the axis and parallel to it. The title should include the names of the variables and succinctly and clearly denote the information on the graph. The reader should not have to read the text to determine the reason for the graph. Place the title so it does not interfere with reading the graph itself.

EXAMPLE
6.4

A thermometer has readings in degrees Celsius; plot the data with degrees Celsius as the ordinate.

Degrees Celsius	Degrees Fahrenheit
0	32
16.7	62
33.3	92
50	122
66.7	152
83.3	182
100	212

SOLUTION:

The graph paper available has five divisions per inch and the distance available for the ordinate is five inches and for the abscissa seven inches. For the abscissa the scale is $212/7 = 30.3°$/inch. In this case round the value to $30°$, which is divisible by five, yielding $6°$ per division. Notice that the entire range for the abscissa data was used to determine the scale, even though the lowest value is $32°$. For the ordinate the scale is $100/5 = 20°$/inch. In this instance the scale fits the division size easily with no adjustment necessary, yielding $4°$ per division. The graph is shown in Figure 6.5. Often it is desirable to combine several graphs on the same chart, as Figure 6.6 illustrates. This is particularly useful in comparative analyses.

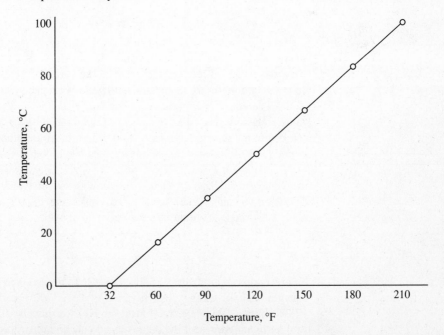

FIGURE 6.5 A line graph of the relationship of degrees Celsius to degrees Fahrenheit.

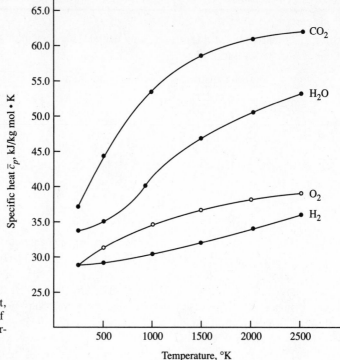

FIGURE 6.6 Multiple line graphs on the same chart, depicting the variation of specific heat with temperature for four gases or vapors.

Equation of a Straight Line

Very often you will want to determine the equation that a set of data yields, to show the empirical relationship between the variables. This is useful when you do not want to refer to the data table to get an ordinate value, given the abscissa value. The equation for a straight line is $y = mx + b$, where y is the ordinate value, x the abscissa value, m the slope, and b the y-intercept when x is zero. We can determine the equation of a straight line from the graphical information in Figure 6.5 as follows (a more sophisticated methodology is described in Chapter 7). The equation we use is $C = mF + b$, where m and b need to be determined from the graph. There was two unknowns, so you pick any two points to determine m and b. Let us pick the following coordinates, (122, 50) and (212, 100), and substitute into the equation above.

$$50 = 122m + b$$
$$100 = 212m + b$$

Subtract one from the other, yielding $m = 0.5556$, and then substitute into either equation and solve for b, which yields $b = -17.78$. Finally, $C = 0.5556F - 17.78$. If the data were not linear, as we determined visually, then using a straight line to represent the data is inaccurate and misleading. For instance, the data for the boiling temperature versus pressure of water are nonlinear.

E X A M P L E

6.5

Determine the plot of the boiling temperature of water versus its pressure for the following data.

Temperature (°C)	Pressure (kilopascals)
0	0.61
50	12.3
100	101.3
150	475.8
200	1553.8
250	3973

SOLUTION:

Determine the scale for the independent and dependent variables. In this case the temperature is the dependent variable. The graph paper has five divisions per inch, and there are five inches available for the ordinate: scale = 250/5 = 50°/inch. This yields 10° per division, as no adjustment is necessary in the scale size. For the pressure the scale is 3973/8 = 496.6 kPa/inch, where there are eight inches available for the abscissa. In this case, round up the scale value to 500 kPa/inch, yielding 100 kPa/division. This is a large value, considering some of the data. A plot of the data is shown in Figure 6.7. The data are nonlinear, and it is difficult using the graph to determine the temperature at low values of pressure. The next section shows a solution of this problem.

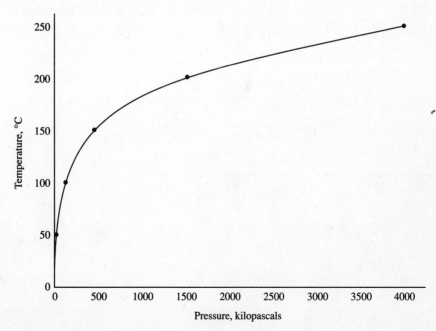

FIGURE 6.7 A line graph of the variation with pressure of boiling temperature of water.

6.4 PLOTTING ON SEMILOG AND LOG-LOG PAPER

A variety of scales other than linear ones can be used for plotting data. Two of the most common involve the use of logarithms. Reasons for considering these are that the data may cover a large range and need to be compressed graphically, and a logarithmic plot may arrange nonlinear data in a straight line. Let's see why the log function causes the data to be compressed.

Semilog and log-log paper is based on logarithms to the base ten, rather than the natural logarithms that are more commonly used in engineering. A logarithm expresses any number by raising the number ten to a power. For instance, take the logs of the pressure data in the previous example.

Temperature (°C)	log (pressure kPa)
0	−0.21
50	1.09
100	2.00
150	2.68
200	3.19
250	3.60

The data are in much more manageable form. The numerical spread of the pressure data is significantly less. Figure 6.8 illustrates a plot of these data.

By using multicycle graph paper, you can extend the range of the data you wish to plot. Figure 6.9a shows a single logarithm scale and Figure 6.9b a three-cycle scale. The three-cycle scale covers a numerical range of one

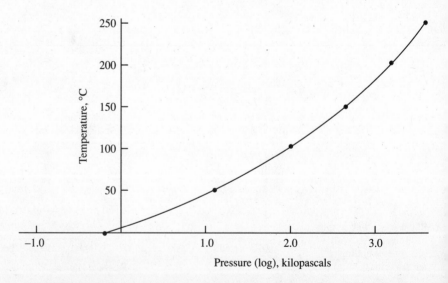

FIGURE 6.8 The plot of the log of pressure versus the boiling temperature for water.

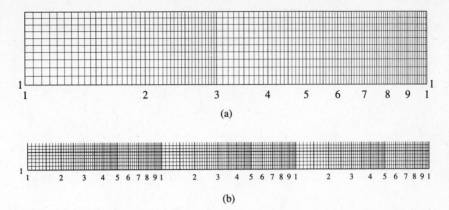

FIGURE 6.9 (a) A single-cycle logarithmic scale. (b) A three-cycle logarithmic scale.

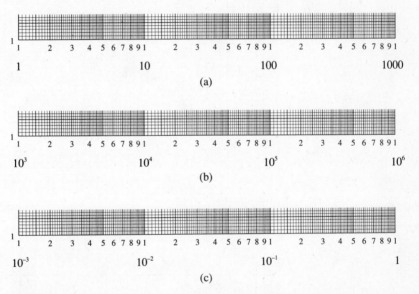

FIGURE 6.10 (a) A three-cycle scale calibrated from 1 to 1000. (b) A three-cycle scale calibrated from 1×10^3 to 1×10^6. (c) A three-cycle scale calibrated from 1×10^{-3} to 1.0.

thousand. Figure 6.10 shows a three-cycle scale calibrated three ways, from 1 to 1000, from 10^3 to 10^6, and from 10^{-3} to 1. When using multiple-cycle graph paper, use the one with the fewest cycles possible to represent your data. The data is most clearly presented in this fashion.

How can you tell if your data should be represented by a semilog or log-log graph, and how does using logarithms produce a straight line from nonlinear data?

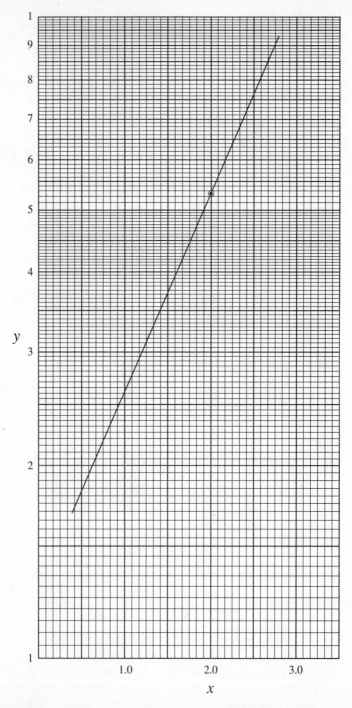

FIGURE 6.11 A graph of the equation $y = 1.3\ e^{0.7x}$ on semilog graph paper.

Consider the following equation:

$$y = be^{mx},\qquad\qquad 6.1$$

and take the log of both sides, yielding

$$\log y = mx(\log e) + \log b.\qquad\qquad 6.2$$

In Equation 6.2 the log b and $m(\log e)$ are constant terms, so the equation is in the form of a straight line, with log y on one side and x on the other. Data that were described by Equation 6.1 would plot as a straight line on semilog paper. One coordinate varies as a log and the other is linear. Figure 6.11 shows this function plotted on semilog paper.

In another common situation, the data are represented by a curve obeying the functional relationship

$$y = bx^{m}.\qquad\qquad 6.3$$

Take the logs of both sides, yielding

$$\log y = m(\log x) + \log b.\qquad\qquad 6.4$$

In Equation 6.4 m and log b are constants, but the two variables are represented as log values. This fits the form of a straight line, so Equation 6.3 is plotted as a straight line on log-log paper. Figure 6.12 illustrates a log-log plot of this function.

EXAMPLE
6.6

Plot the following data on log-log paper and determine the equation relating x and y.

x	y
1	2.0
3	11.6
5	26.3
7	45.0
9	67.3

SOLUTION:

Figure 6.13 illustrates the plot of the data. To find the equation describing the data, substitute data from two points to solve for the equation of a straight line on the log-log plane. Equation 6.4 is used at the two points.

$$\log 26.3 = m(\log 5) + \log b$$
$$\log 67.3 = m(\log 9) + \log b$$

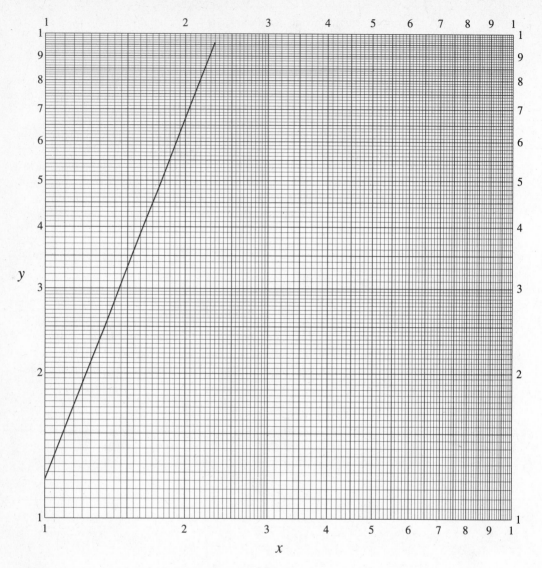

FIGURE 6.12 A log-log plot of the equation $y = 1.2x^{2.5}$.

This yields upon evaluating the logs

$$1.420 = 0.699m + \log b$$

and

$$1.828 = 0.954m + \log b.$$

Subtract the first from the second, eliminating the log b.

$$0.408 = 0.255m$$
$$m = 1.6$$

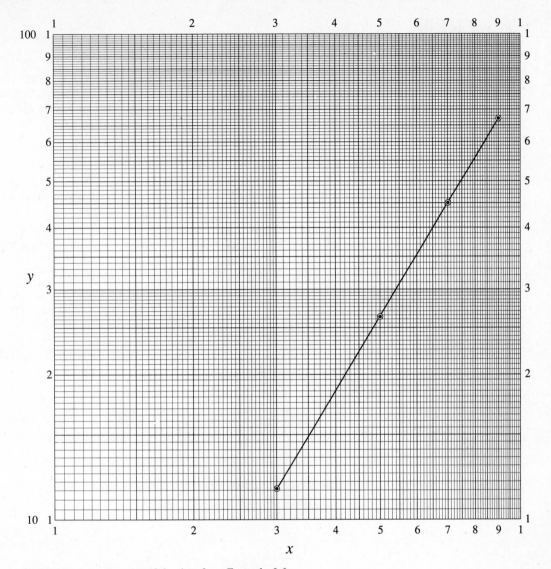

FIGURE 6.13 A log-log plot of the data from Example 6.6.

Substitute the value of m in either equation and solve for b.

$$1.42 = 0.699 \times 1.6 + \log b$$
$$\log b = 0.3016$$
$$b = 2.00$$

The equation is $y = 2.0x^{1.6}$. To verify that this is true, check the equation at another point, $y = 2.0(3.0)^{1.6} = 11.6$.

■

6.5 SKETCHING

A sketch is not a sloppy drawing, nor is it inaccurate; rather it is an accurate, freehand drawing used to transmit information visually. To sketch well you need, in addition to the techniques presented here, the ability to visualize an object. A course in engineering graphics or engineering drawing is very helpful in this regard.

The only equipment you need are a pen or pencil and a sheet of paper. When you want to sketch in pencil, a medium-weight pencil is best. You want one that will not smudge but is not so hard it prevents you from varying line intensity and width. Most often you will sketch a device so it looks three-dimensional. An isometric pictorial drawing is particularly useful, and we will cover that briefly. It is convenient, though not a requirement, to use paper printed with an isometric grid pattern, shown in Figure 6.14. This paper has lines at a 30° angle to the horizontal, which help give the picture depth.

Let's consider the techniques used to draw an elementary sketch.

STEP 1. Lay out the three-dimensional grid, keeping lines parallel and in this case vertical lines vertical, but with what would be the *x* and *y* coordinates slanted at 30° (Figure 6.15a). It is possible to use 45° as well, though the slope is a little dramatic. The construction lines should be light and thin, as you will erase them when the drawing is completed. The model should fit entirely within this frame.

STEP 2. Draw the object you have in mind, remembering the parallel rule for lines. Also, circles are not circles, but ellipses in isometric sketches (Figure 6.15b).

STEP 3. Remove the construction lines. Some feel this is not a necessary step, but the visual difference between Figures 6.15b and 6.15c seems important.

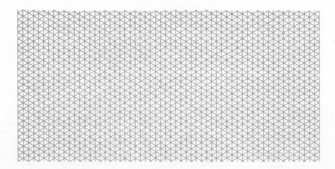

FIGURE 6.14 An isometric grid pattern.

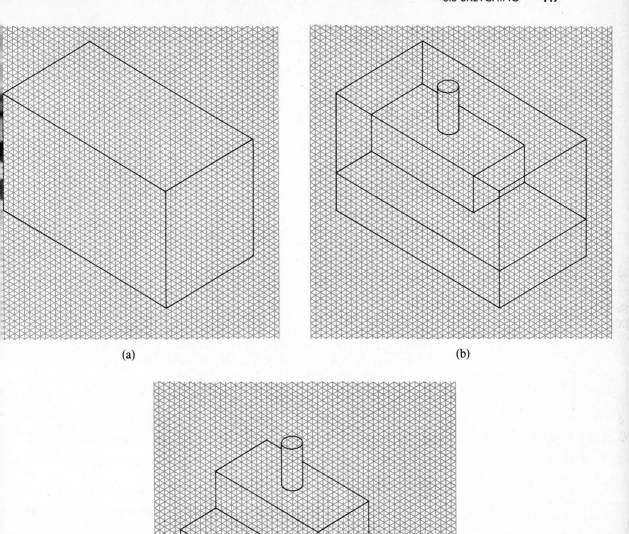

(a)   (b)

(c)

FIGURE 6.15 (a) Lay out the grid with construction lines (light and thin). (b) Draw the lines of the object. (c) Remove the construction lines.

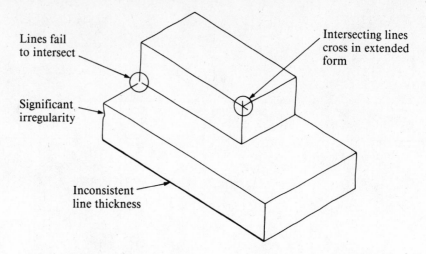

Lines fail
to intersect

Intersecting lines
cross in extended
form

Significant
irregularity

Inconsistent
line thickness

FIGURE 6.16 Sketching errors.

Some additional errors that detract from the appearance of your sketch are shown in Figure 6.16. Of course the errors are exaggerated, but they certainly have a negative impact. Have all lines intersect and terminate where they should, not forming crosses or other shapes that are distracting and, in the worst situation, misleading. Equally bad is to have the lines not meet at all, or some of the lines not meet with a common terminus. Sketches are freehand drawings, so some irregularity in line direction is expected, but significant irregularities should be corrected. Visible lines should be dark and of uniform thickness: this may mean sharpening your pencil occasionally. The thick line shown in Figure 6.16 detracts from the sketch.

REFERENCES

1. Earle, J. H. *Engineering Design Graphics.* 4th ed. Addison-Wesley, Reading, MA, 1983.
2. Spence, W. P. *Engineering Graphics.* Prentice Hall, Englewood Cliffs, NJ, 1985.
3. Voland, G. G. S. *Modern Engineering Graphics and Design.* West, St. Paul, 1987.

PROBLEMS

6.1 Construct a circle chart depicting the following grade distribution data for a class: 7 A's, 12 B's, 21 C's, 10 D's, and 4 F's.

6.2 Construct a circle chart indicating the distribution of engineering majors at a university given the following information: chemical engineering—210 students; civil—355; electrical—689; industrial—310; and mechanical—474.

6.3 Construct a circle chart illustrating a state's annual expenditure on public roads. Approximate annual figures are construction—$14 million; maintenance—$8 million; equipment purchase and maintenance—$3.5 million; bonds—$6.2 million; engineering and administration—$2.1 million.

6.4 An alloy has the following composition: lead—24%; tin—16%; aluminum—16.6%; copper—4.6%; zinc—38.8%. Create a circle chart for the alloy.

6.5 A company has the following age distribution among its employees:

Age group	<20	20–29	30–39	40–49	50–59	>59
Employees	15	71	82	68	51	19

Construct a circle chart showing the distribution by age of the employees.

6.6 Construct a bar chart for the data in problem 6.1.

6.7 Construct a bar chart for the data in problem 6.2.

6.8 Construct a bar chart for the data in problem 6.3.

6.9 Construct a bar chart for the data in problem 6.4.

6.10 Construct a bar chart for the data in problem 6.5.

6.11 A manufacturing company has the following annual expenditures: research and development—$21 million; engineering design—$11 million; tooling and equipment—$15 million; labor—$33 million; and marketing/distribution—$20 million. Construct a bar chart illustrating these expenditures.

6.12 A manufacturing company produces two models of a milling machine and has the following weekly production forecast.

Year	Model A	Model B
1989	550	700
1990	600	825
1991	650	850
1992	500	950
1993	400	1100
1994	0	1800

Construct one bar chart showing the weekly production forecast for both models.

6.13 The following data were obtained from a laboratory experiment where a structure was subjected to a force, causing it to deflect.

Force (kN)	Deflection (mm)
0	0
100	0.3
300	1.8
500	2.9
700	4.1
900	5.2
1100	6.3

Construct a line graph of the data and determine the equation of a straight line passing through the points.

6.14 The following data represent the weight per foot of wire rope for various wire diameters. Construct a graph of the data.

Diameter (in)	Weight (lbf/ft)
¾	1.41
1	2.50
1¼	3.91
1½	5.63
1¾	7.66
2	10.00
2¼	12.50
2½	15.2
2¾	18.3
3	22.2
3½	29.9
4	38.4

6.15 The following data represent the standard temperature and pressure variation with altitude. Plot both temperature and pressure variation on the chart.

Altitude (ft)	Temperature (°F)	Pressure (psia)
0	59.0	14.69
5,000	41.2	12.22
10,000	23.3	10.10
15,000	5.5	8.29
20,000	−24.6	6.75
25,000	−30.2	5.45
30,000	−48.0	4.36
35,000	−65.8	3.46
36,089	−69.7	3.28
40,000	−69.7	2.72
50,000	−69.7	1.68

6.16 A diesel engine has the following horsepower outputs as a function of a percentage of its rated rpm. Two outputs are given for 80% throttle and 100% throttle. Plot both curves on the same chart.

Percentage-rated rpm	80% (hp)	100% (hp)
40	30	40
60	52	66
80	70	86
100	80	100

6.17 The bending moment of a shaft supported by bearings is

$$M = wL^2/12,$$

where M is the bending moment in inch-lbf, L is the span between bearings in inches, and w is the weight of a unit length of shaft, lbf/inch. Plot the bending moment for a shaft with $w = 15$ lbf/inch for the span varying from 6 inches to 2 feet in 2-inch intervals.

6.18 A capacitor is discharged and the following table gives the values of the charge in coulombs in the capacitor with time. Plot this data.

Time	Charge (C)
0	25×10^{-6}
0.1	16.8×10^{-6}
0.2	11.3×10^{-6}
0.3	7.5×10^{-6}
0.4	5.0×10^{-6}
0.5	3.5×10^{-6}
0.6	2.3×10^{-6}
0.7	1.5×10^{-6}
0.8	1.0×10^{-6}
0.9	0.8×10^{-6}

6.19 The following table indicates the temperature-time history of a metal ingot when it was immersed in boiling water. Plot the information and determine the equation of a straight line describing the data points.

Time (sec)	Temperature (°F)
0	-10
5	5
10	20
15	35
20	50
25	65
30	80

6.20 A group of machine parts, made of several components, were tested repetitively under a variety of loads until they failed. The following data indicate the loads and the cycles required until failure occurred. Plot the data on semilog paper and determine the equation relating cycles to failure as a function of load.

Load (lbf)	N (cycles to failure)
10	4.10×10^6
15	2.74×10^6
20	1.96×10^6
25	1.40×10^6
30	1.00×10^6
35	7.18×10^5
40	5.14×10^5

6.21 For the data presented in problem 6.14, determine the log-log plot and the equation describing the weight in terms of the diameter.

6.22 Plot the pressure-altitude data from problem 6.15 on semilog paper and determine the equation for pressure as a function of altitude.

6.23 Plot the data from problem 6.17 on log-log paper.

6.24 Plot the data from problem 6.18 on semilog paper and determine the equation for charge remaining as a function of time.

6.25 Plot the function $y = 1.9 \, x^{2.1}$ on rectangular graph paper and on log-log paper.

6.26 Plot the function $y = 1.9 \, x^{-2.1}$ on rectangular graph paper and on log-log paper.

6.27 Plot the function $y = 2.3 \, e^{1.5x}$ on rectangular graph paper and on semilog paper.

6.28 Plot the function $y = 2.3 \, e^{-1.5x}$ on rectangular graph paper and on semilog paper.

6.29 A gear set in a laboratory can lift a certain amount of weight, given in the following table. Determine the equation that best describes the data.

Applied force (lbf)	Weight lifted (lbf)
15	184
29.3	286.4
35.4	334.1
46.4	375
55.9	463.6

6.30 Determine the equation that best describes the data.

X	Y
1	3
2	2.27
3	1.93
4	1.72
5	1.57
6	1.46

6.31 Determine the empirical equation that best fits the data.

X	Y
1	80.3
1.5	270
2.0	1,614
2.5	7,232
3.0	32,412

6.32 Draw an isometric sketch of a two-inch cube five times. Each version should have one of the defects, only once, as shown in Figure 6.16, and one should be correct. Look at the sketches and see how the defect detracts from the drawing, hence from your communication using the sketch.

6.33 Draw sketches (a) through (f) yourself. Notice that it is much easier to sketch when the grid lines of the graph are included.

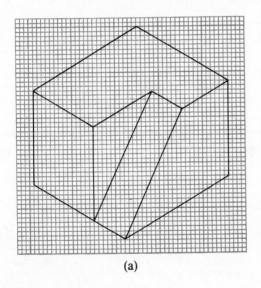

(a)

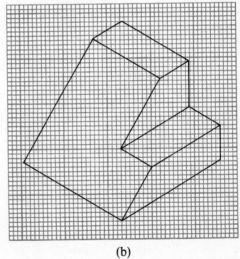

(b)

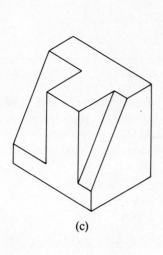

(c)

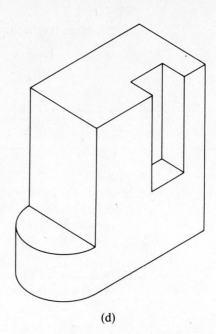

(d)

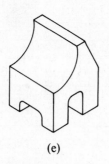

(e)

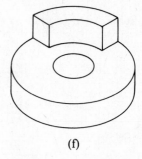

(f)

STATISTICS AND ERROR ANALYSIS

- To learn the measures of central tendency of data.

- To learn the measures of data dispersion.

- To become familiar with probability theory and normal distributions.

- To apply linear regression analysis.

- To understand error analysis and error propagation.

(USDA/SCS Photo)

Engineering students are often involved with statistics as part of their educational program, sometimes in course work, frequently in labs. Engineers in industry use statistics and the results of statistical analysis as part of their work function.

7.1 INTRODUCTION

Statistics is the collection, organization, and interpretation of numerical data. The data may be from experiments run in a laboratory or from economic information. Daily we are deluged with a variety of statistical information, ranging from the weather forecasts, to inflation rates, to the predicted outcome

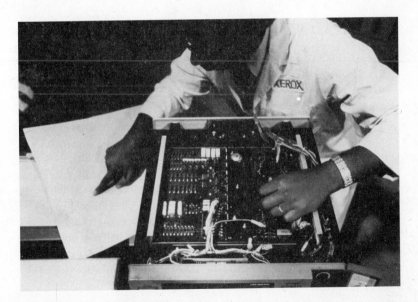

This engineer is checking performance to make sure it meets the design operating specifications. In a quality assurance program a certain number of units are randomly sampled and checked for their performance. This subgroup has been mathematically shown to be representative of the whole and is used as a predictor of the group quality. (Photo Courtesy of Xerox Corporation.)

of political elections. It is important that we be able to interpret this information wisely as statistics can also easily befuddle us at times.

In industry you will use statistical analysis in a variety of circumstances. For instance, in a manufacturing plant the quality control engineer must sample and test a certain number of the components to assure that the manufacturing process and product are meeting the expected quality standards. Then the engineer must judge from the measurements on a few whether the majority are within bounds. Obviously, this is a complex area of study; people spend their entire careers specializing in statistics, and we cannot hope to delve into all the complexities of statistics here. However, an understanding of fundamentals will help you perform better as engineering students and as practicing engineers.

7.2 MEASURES OF CENTRAL TENDENCY

When we have collected a set of numerical data, we most often want to characterize it so we can visualize the cumulative nature of data, rather than the entire set. Most frequently we look at the average value of the data, the central tendency of the data. The average that we most often have in mind is the arithmetic mean, $\bar{x}$, which is defined as

$$\bar{x} = \sum_{i=1}^{n} \frac{x_i}{n}$$ 7.1

where $\bar{x}$ is the average value, x_i is the individual values that are to be averaged, and n is the total number of x_i's. The symbol Σ represents summation. Let's see how this is used. Find the arithmetic mean of the numbers 10, 8, 12, 5, 7, 9, and 9.

$$\bar{x} = \sum_{i=1}^{7} \frac{x_i}{n} = \frac{10 + 8 + 12 + 5 + 7 + 9 + 9}{7} = \frac{60}{7} = 8.6$$

The mean gives us an understanding of the centrality of the available data group, but the results can be misleading if there are some data that are not characteristic of the group. In that case the median is better used to determine the centrality. The median is found by arranging the data in order of magnitude from smallest to largest. The value at the halfway point, if the number of data is odd, is the median value. If the number of data is even, then the average of the middle two values is the median value of the data.

■
E X A M P L E
7.1

The following data represent the prices of new automobiles purchased by your neighbors in the past two years. Determine the mean and median values: $15,000, $17,000, $11,000, $13,000, $43,000, $15,000, $12,000.

SOLUTION: The mean value is

$$\frac{126,000}{7} = \$18,000.$$

Arranging the data in ascending value yields $11,000, $12,000, $13,000, $15,000, $15,000, $17,000, $43,000. The median value is $15,000. In this case, because one car cost $43,000, an anomaly compared to the rest of the data, the median is more indicative of the centrality of the data.

■

There are other measures of the central tendency of the data. The mode can be useful at times, particularly when looking at data distributions. A mode occurs when a piece of data repeats itself; for instance, in the car price data, the value of $15,000 occurred twice, so the data have one mode. If no value is repeated, the data have no modality, and if different values repeat, then there are two, three, or more modes associated with the data.

The following numbers have the same mean and median, but different modes:

2, 6, 4, 3, 9 no modes

7, 7, 4, 3, 3 two modes (bimodal)

We do not have to adopt just one measure but can use our common sense and multiple measures to characterize the data.

A final measure that we will analyze is the weighted arithmetic mean, $\overline{X}$, which is used when certain data values have more importance (weight) than others. We use weighting factors in Chapter 7 in discussing multicriteria decision making. In this case the weighting factors are w_1, w_2, w_n, . . . , and are associated with numbers X_1, X_2, X_n, ... The weighted arithmetic mean is

$$\overline{X} = \frac{w_1X_1 + w_2X_2 + w_3X_3 + \cdots + w_nX_n}{w_1 + w_2 + w_3 + \cdots + w_n} = \sum_{i=1}^{n} \frac{w_iX_i}{\Sigma w_i}. \qquad 7.2$$

■
E X A M P L E
7.2

Given the following data with their weighting factors, determine the weighted arithmetic mean.

Data (X)	Weighting factor (w)
10	1
12	1.5
20	1
15	2
19	3
16	1

SOLUTION: The weighted arithmetic mean is

$$\overline{X} = \frac{1 \times 10 + 1.5 \times 12 + 1 \times 20 + 2 \times 15 + 3 \times 19 + 1 \times 16}{1 + 1.5 + 1 + 2 + 3 + 1} = 15.9.$$

∎

7.3 MEASURES OF DISPERSION

Just as a set of numbers has a centrality, so it has a dispersion. The scattering of the data about a central point provides us with additional information about the set. Consider the number set (10, 10, 10, 10, 10), which has a mean of 10, and another set with the same mean, (1, 20, 14, 5, 10). The mean does not indicate the scattering of the second group of data.

Several indicators can be used to indicate the scattering of the data. The range, R, is defined as the difference between the largest and smallest values of the data. Thus, for the two data sets above the range of the first set is 0 and of the second set 19.

An individual deviation is the difference between any data point and the mean. Note that the sum of the individual deviations is zero: the individual deviations may be positive or negative, depending on whether or not any given data point is greater than or less than the mean, and since they are equally dispersed about the mean their sum is zero.

A composite deviation would be useful, and there are ways to overcome the problem of the deviations' sum being zero. We could use the absolute value of the deviations and sum these: this is the mean deviation. A preferred method is the standard deviation. The standard deviation is also used with the normal distribution curve in probability theory (Section 7.4) and hence becomes the most common and useful dispersion correlator.

The standard deviation, σ, is defined as

$$\sigma = \sqrt{\sum_{i=1}^{n} \frac{(x_i - \overline{x})^2}{n}}$$ 7.3

∎

E X A M P L E
7.3

For the following set of numbers determine the range and the standard deviation: 5, 8, 10, 15, 12, 18, 11, 7.

SOLUTION: The range is $R = 18 - 5 = 13$. The mean is

$$\overline{x} = \sum_{i=1}^{n} x_i/n = 86/8 = 10.75.$$

x_i	$(x_i - \bar{x})^2$
5	33.06
8	7.56
10	0.56
15	18.06
12	1.56
18	52.56
11	0.06
7	14.06

$$\sum_{i=1}^{n} (x_i - \bar{x})^2 = 127.48$$

$$\sigma = \sqrt{127.48/8} = 3.99$$

■

7.4 PROBABILITY AND NORMAL DISTRIBUTION

Very often when we think of statistics and probability we think chances: what is the chance that something will happen? Probability helps in estimating chances. Consider flipping a coin: what is the chance, the probability, that it will be heads when it lands? The probability of an event occurring is defined as the frequency of the occurrence divided by the number of attempts, letting the number of attempts be quite large. In engineering probability is important in sampling of parts for quality control; the quality of the sampled parts should relate to that of the remaining parts, and should be the same if the sample is properly selected. Thus probability theory and statistics are closely allied.

Returning to the coin, the probability is

$$p = s/n = 1/2, \qquad 7.4$$

where s is the number of desired outcome (in this case there is one outcome, heads), and n is the number of possible outcomes of an event (in this case two, heads or tails). The probability of all other events, q, is

$$q = (n - s)/n. \qquad 7.5$$

If we combine Equations 7.4 and 7.5 we find that the sum is one, or 100 percent probability that some event will occur. Thus, when the probability is one, the event is certain to occur, and when the probability is zero the event cannot occur.

Any time you flip a coin the probability is 50 percent that it will come up heads, regardless of how many times you have flipped it before. Say you are matching coins with a friend. You both want to have heads occur simultaneously. The probability for you getting heads is 1/2, the probability for your friend is 1/2, and the combined probability is the product of the two, or 1/4, 25 percent. To generalize this as a law, if the probability of m independent events is p_1, p_2, p_3, . . ., p_m, then the probability that they will all occur simultaneously, P, is

$$P = p_1 p_2 p_3 \cdots p_m. \qquad 7.6$$

Now let's consider another situation. You have neatly placed ten pennies on a table, near the edge, heads up. Your younger brother comes along and sweeps them off the table. What is the probability that they will all be heads up when they land on the floor? $P = (1/2)^{10} = 1/1024$, or about one chance in a thousand. Let's change the situation to that of at least one being heads up. In this case the law in Equation 7.6 no longer applies, as we are no longer asking that all looked-for events happen simultaneously. The probability, P, that one of the looked-for events will occur when there are m independent events occurring with their individual probabilities is

$$P = p_1 + p_2 + p_3 + \cdots + p_m. \qquad 7.7$$

For our situation the individual probabilities are 1/2, so the probability that at least one coin will be heads up on the floor is $P = 10(1/2) = 5/1$. In both of these cases the events must be mutually independent: one event does not affect the other events. If it did, then the probability function cannot be as readily determined, and Equations 7.6 and 7.7 are not valid.

The frequency distribution of an event is often important in laboratory situations when you measure a quantity that is supposedly constant, but actually varies somewhat.

E X A M P L E

7.4

An engineer is gathering data on a power plant. At a certain place in the power plant the steam pressure is supposedly constant, but actually varies slightly. There were 100 observations, yielding the following results.

Pressure (psia)	Number of results (n)
397	1
398	3
399	11
400	26
401	33
402	17
403	6
404	3

SOLUTION:

These data are plotted in Figure 7.1a, a histogram. This gives a frequency distribution of the data.

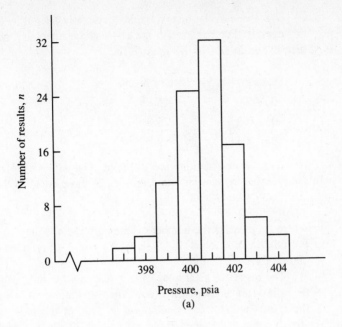

(a)

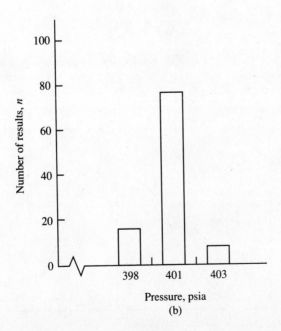

(b)

FIGURE 7.1 (a) A histogram for the data in Example 7.4. (b) A histogram for consolidated data in Example 7.4.

The shape of the histogram can be greatly altered by changing the number of intervals. For instance, if we chose three intervals, the data would have a distribution as shown in Figure 7.1b.

Pressure	Number of results
398–400	15
401–402	76
403–404	9

Clearly, the implications are different. Guidelines help in selecting the best interval size for plotting the data. One of these is called the Sturgis rule:

$$I = 1 + 3.3 \, (\log n), \qquad 7.8$$

where I represents the number of intervals and n is the number of observations. For this situation $I = 1 + 3.3 \, (\log 100) = 7.6$, or rounded off to 8, the number chosen for the example.

The results of Example 7.4 still may not be quite what you have in mind regarding frequency distribution. When the number of observations is infinite, the results will form a smooth curve. One of the most common distribution curves is the bell-shaped curve, or normal distribution curve, shown in Figure 7.2. (Discussion of other distribution curves is beyond our scope.) The equation describing the curve in Figure 7.2 is

$$Y = \frac{1}{\sigma\sqrt{2\pi}} \left[e^{-(x-m)^2/2\sigma^2} \right], \qquad 7.9$$

where Y is the number of results (the frequency), x is the magnitude of the measured event (the pressure), m is the mean value of the events, and σ is the standard deviation.

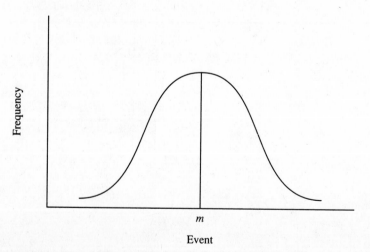

FIGURE 7.2 A normal distribution curve.

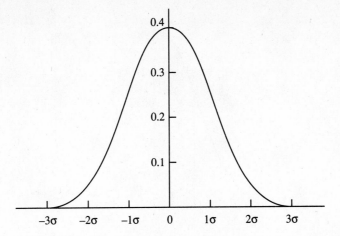

FIGURE 7.3 A standardized normal distribution curve.

This is not the typical fashion in which the normal, or Gaussian, distribution curve is given, as the variables may be changed to make the standard normal distribution curve. To do this let

$$y = Y\sigma \quad \text{and} \quad z = (x - m)/\sigma. \qquad 7.10$$

Substitution into Equation 7.9 yields

$$y = \frac{1}{\sqrt{2\pi}} (e^{-z^2/2}). \qquad 7.11$$

A plot of Equation 7.11 is shown in Figure 7.3. A tremendous advantage to using this curve is that the abscissa is in multiples of the standard deviation. Furthermore, the area under the curve integrates to unity, or 100 percent probability. The following table, calculated from Equation 7.10, shows why the abscissa is in increments of the standard deviation.

x	z
$m + 2\sigma$	$+2$
$m + 1\sigma$	$+1$
m	0
$m - 1\sigma$	-1
$m - 2\sigma$	-2

Figure 7.4a shows that 68.3 percent of the area, or data, lies within plus or minus one standard deviation of the mean ($z = 0$); Figure 7.4b that 95.5 percent of the area lies within plus or minus two standard deviations of the mean; and Figure 7.4c that 99.7 percent of the area lies within plus or minus three standard deviations of the mean.

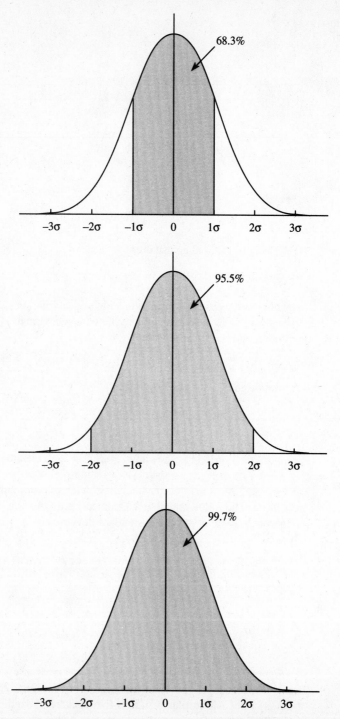

FIGURE 7.4 (a) 68.3% of the data are within $\pm\sigma$ of the mean based on a standardized normal distribution curve. (b) 95.5% of the data are within $\pm 2\sigma$ of the mean based on a standardized normal distribution curve. (c) 99.7% of the data are within $\pm 3\sigma$ of the mean based on a standardized normal distribution curve.

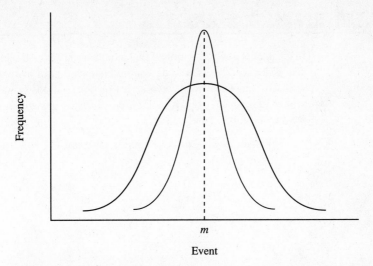

FIGURE 7.5 Two normal distribution curves with the same mean value, but different standard deviations (σ_2, σ_1).

The smaller the standard deviation, the closer the data fit together. Figure 7.5 illustrates this with two normal distribution curves, having the same mean value, m, but looking entirely different. In the narrow distribution curve, the standard deviation has a smaller value than in the broader curve. Thus the probability of being distant from the mean is less.

A note of caution. Not all distributions are normal ones; some may be bimodal, Figure 7.6a, or skewed, Figure 7.6b. These frequently occur in grade distributions, among other situations, and applying a normal distribution function to these may lead to erroneous conclusions. Don't assume the data set you are analyzing follows the Gaussian distribution—plot it to be sure.

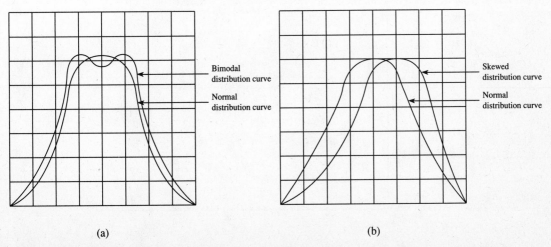

FIGURE 7.6 (a) A bimodal distribution curve and a normal distribution curve. (b) A skewed distribution curve and a normal distribution curve.

7.5 LINEAR REGRESSION ANALYSIS

In Chapter 6 it was mentioned that a more sophisticated means, linear regression analysis, could be used to determine the equation of a straight line. Perhaps the data from an experiment or a series of observations produce, when plotted, a scatter diagram as shown in Figure 7.7. The data are from an experiment on measuring voltage and current.

Voltage (x)	Current (y, milliamperes)
2	12
3	20
4	24
6	35
7	38
8	45
10	53

What we would like to determine is the equation of a straight line that best represents the data.

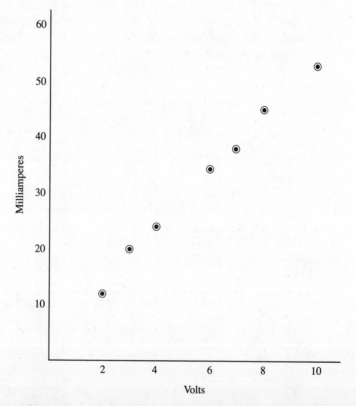

FIGURE 7.7 A scatter diagram of the relationship between volts and milliamperes.

You could take a transparent straight edge and align it as closely as possible with the data, draw a line, and determine the equation of this line by the slope-intercept method. You want the deviations from the line to any point to be a minimum, considering all the data points. It is possible to perform this task mathematically. Figure 7.8 graphs the data with a straight line and shows the deviation, d_i, at one point. The deviation is the difference in y values between the line's value and the data point at a given value of x. What we want to do mathematically is to make the sum of the squares of the deviations a minimum. Squaring the deviations prevents the positive and negative deviations from canceling each other and prevents a line, not fitting the data at all, from meeting the constraint of minimum sum deviation.

The mathematics for determining the coefficients of the line, $y = a + bx$, is not complex but requires differential calculus. The results for any data set (x, y) are given in Equations 7.12 and 7.13.

$$b = \frac{n \sum_i x_i y_i - \sum_i x_i \sum_i y_i}{n \sum_i x_i^2 - (\sum_i x_i)^2}$$

7.12

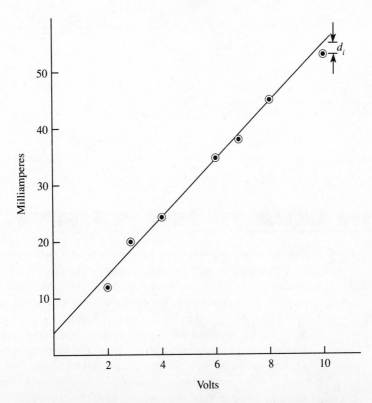

FIGURE 7.8 A straight line, $y = a + bx$, drawn through the data points from Figure 7.7. d_i = deviation of the data from the line at point $x = 10$.

$$a = \frac{\sum_i y_i - b \sum_i x_i}{n} \qquad 7.13$$

The data from Figure 7.7 result in the following values for a and b.

x	y	x^2	xy
2	12	4	24
3	20	9	60
4	24	16	96
6	35	36	210
7	38	49	266
8	45	64	360
10	53	100	530
$\sum_i x_i = 40$	$\sum_i y_i = 227$	$\sum_i x_i^2 = 278$	$\sum_i x_i y_i = 1546$

$$b = \frac{(7)(1546) - (40)(227)}{(7)(278) - (40)^2} = \frac{1742}{346} = 5.035$$

$$a = \frac{(227) - (5.035)(40)}{7} = \frac{25.6}{7} = 3.657$$

The equation of the straight line is $y = 3.657 + 5.035x$.

7.6 ERROR ANALYSIS AND ERROR PROPAGATION

We have been assuming that the measurements, with resulting data, are correct in our problems and examples. Any readings that are made involve error, however. Systematic errors are associated with the instrument or the technique used in the measurement. For instance, a thermometer reads 102°C rather than 100°C for the boiling point of water. A yardstick is not exactly 36 inches long, and the inch markings are not exactly one inch. In looking at a gage you must interpret where the needle is pointing; a systematic error is associated with reading the value from the instrument. Random errors are produced by a wide variety of unpredictable variations in the experiment. Perhaps fluctuations in atmospheric pressure or opening and closing the laboratory door causes variations in the experimental results. These random errors cannot be duplicated.

A third kind of error is not an error at all but a blunder. Perhaps you made a mistake in reading an instrument or didn't calibrate a piece of test apparatus, and the results are in error. Ideally, the experiment should be rerun.

The terms *accuracy* and *precision* refer to the two types of errors that will occur in any experiment. High accuracy refers to a measurement that has a small systematic error. A micrometer, measuring to one-thousandth of an inch, is used to measure thickness, rather than a ruler that measures to the nearest thirty-second of an inch. An experiment with high precision is one in which the random errors are small, and the experiment is repeatable. The experimental design should minimize errors, by controlling extraneous events to minimize random errors and by using accurate instrumentation and measurement techniques to minimize systematic errors. In general you will not have control over the instrumentation that is used in experiments in engineering and physics, but you can be aware of the errors that are created by use of the existing instrumentation.

The most important aspect of error analysis is calculating the effect that individual errors in experimental observation have on the final result. Thus, by knowing the error in a temperature reading we can determine the error in a calculation using the temperature. The percent error in any reading is the uncertainty of the reading divided by its nominal value. A pipe's diameter is measured and found to be 4 centimeters, within 0.008 centimeters, so the percent error associated with the measurement is $(0.008/4) \times 100 = 0.2\%$. Consider that you are measuring a cylinder's volume by determining its diameter and height and then calculating the volume. Let Δr and Δh represent the error in the measurement of the radius, r, and height, h. The initial volume is

$$V = \pi r^2 h,$$

The control room in a nuclear power plant provides the operators with information concerning all aspects of the plant operating condition. (Courtesy of Trident Engineering Associates)

and the volume with the measurement errors is

$$V + \Delta V = \pi(r + \Delta r)^2(h + \Delta h). \tag{7.14}$$

Thus,

$$V + \Delta V - V = \pi(r^2\Delta h + 2rh\Delta r + \Delta r^2 h + 2r\Delta r\Delta h + \Delta r^2\Delta h); \tag{7.15}$$

neglecting higher-order terms as being too small compared to the remaining terms, Equation 7.15 becomes

$$\Delta V = \pi(r^2\Delta h + 2rh\Delta r). \tag{7.16}$$

If we divide by V to obtain the error in the total volume, the result is

$$\Delta V/V = \frac{\pi(r^2\Delta h + 2rh\Delta r)}{\pi r^2 h} = \frac{\Delta h}{h} + \frac{2\Delta r}{r}. \tag{7.17}$$

For instance, if the radius is 50 cm and $\Delta r = 0.5$ cm, and the height is 100 cm and $\Delta h = 0.5$ cm, then the percent error in the volume is

$$\frac{\Delta V}{V} = \frac{0.5}{100} + \frac{(2)(0.5)}{50} = 0.025 = 2.5\%.$$

This can be extended to more complicated equations. I will not derive the general error equation, but will demonstrate its use in the following example. Let us determine the percent error in Y, given by the equation below,

$$Y = AB^m/C^n, \tag{7.18}$$

where A, B, C are measured quantities and m and n are the exponents of these variables. The percent error in Y is

$$\Delta Y/Y = \sqrt{\left(\frac{\Delta A}{A}\right)^2 + \left(\frac{m\Delta B}{B}\right)^2 + \left(\frac{n\Delta C}{C}\right)^2}. \tag{7.19}$$

Notice that the effect of the negative sign on n is not important, as the value of $\Delta C/C$ is squared.

E X A M P L E
7.5

The period of a pendulum is given by the equation $t = 2\pi(L/g)^{1/2}$. In a physics lab the following measurements are made: $t = 2$ seconds within 0.02 seconds; the length of the pendulum is 1 meter within 0.01 meter. Determine the percent error in the gravitational acceleration, and the values of the gravitational acceleration and the error.

SOLUTION:

$$g = 4\pi^2 L/t^2$$

$$\Delta g/g = \sqrt{\left(\frac{2\Delta t}{t}\right)^2 + \left(\frac{\Delta L}{L}\right)^2} = \sqrt{\left(\frac{(2)(0.02)}{2}\right)^2 + \left(\frac{0.01}{1}\right)^2}$$

$$\Delta g/g = 0.022 = 2.2\%$$

$$g = \frac{4\pi(1)}{2^2} = 9.869 \text{ m/s}^2$$

$$\Delta g = 0.022 \ (9.869) = 0.217 \text{ m/s}^2$$

■

REFERENCES

1. Beckwith, T. G.; Buck, N. L.; and Marangoni, R. D. *Mechanical Measurements*. 3d ed. Addison-Wesley, Reading, MA, 1982.
2. Neville, A. M., and Kennedy, J. B. *Basic Statistical Methods for Engineers and Scientists*. International Textbook, Scranton, PA, 1964.
3. Walpole, R. E., and Myers, R. H. *Probability and Statistics for Engineers and Scientists*. 4th ed. Macmillan, New York, 1989.
4. Young, H. D. *Statistical Treatment of Experimental Data*. McGraw-Hill, New York, 1962.

PROBLEMS

7.1 Determine the range, mean, and median for the following numbers: 9, 7, 12, 13, 10, 9, 8.

7.2 Determine the range, mean, and median for the following numbers: 15, 16, 12, 17, 15, 10, 11.

7.3 The following numbers represent the test scores of a group of students: 82, 75, 64, 86, 77, 81, 68, 94, 93, 65, 49, 77, 64, 55. Determine the arithmetic mean, median, and mode for the grades.

7.4 The chief engineer at a small manufacturing plant is determining the average monthly utility bill for the operations. The monthly bills are $200, $245, $175, $325, $210, $390, $295, $215, and $255. Determine the mean and median.

7.5 The number of cars crossing an intersection is measured for 15-minute intervals at various times of day, yielding the following data.

Time of day	Cars per 15 minutes
12 midnight to 6:00 A.M.	10
6:00 A.M. to 10:00 A.M.	78
10:00 A.M. to 1:00 P.M.	45
1:00 P.M. to 4:00 P.M.	40
4:00 P.M. to 8:00 P.M.	84
8:00 P.M. to 12 midnight	32

Determine the mean and median in units of cars per hour.

7.6   The rate of water flow to a processing plant varies with the time of day. The hourly rate data in gallons per minute (gpm) are 250, 200, 205, 290, 380, 500, 490, 400, 450, 500, 420, 450, 390, 380, 425, 505, 530, 500, 440, 400, 350, 300, 300, 290. Determine the mean, median, and mode.

7.7   A small machine shop has an inventory of galvanized metal sheeting, the numbers represent the size of each sheet: 8, 8, 9, 9, 9, 10, 10, 11, 12, 13, 14, 14, 14, 15, 16, 17, 17, 19. Determine the mean, median, and mode. How would you best represent the group with these centrality indices, bearing in mind that there are no fractional sizes?

7.8   Two sets of data have mean values of 30, but set A has a standard deviation of 2.0 and set B a standard deviation of 5.0. Sketch the differences between the two sets of data.

7.9   Using the data set from Problem 7.7 and letting the number of occurrences of a number be its weighting factor, determine the weighted average for the group.

7.10   The specific heat of a gas mixture is found by using the weighted average of the individual gas-specific heats. The following table lists the gases, weighting factors, and specific heats.

Gas	Weighting factor	Specific heat (kJ/kg-K)
Carbon dioxide	0.30	0.844
Sulfur dioxide	0.30	0.6225
Helium	0.20	5.1954
Nitrogen	0.20	1.0399

Determine the average specific heat of the mixture (the weighted average of the specific heats).

7.11   Determine the standard deviation for the data in Problems 7.1 and 7.2.

7.12   Determine the standard deviation for the data in Problem 7.3.

7.13   Determine the standard deviation for the data in Problem 7.6.

7.14   Write a computer program that will read data and calculate the arithmetic mean and the standard deviation.

7.15   Using the data in Problem 7.4, determine the standard deviation and plot the normal distribution curve.

7.16   In a manufacturing plant a steel rod is supposed to be cut to a length of 7.80 inches; however, measurements of the steel rods that were cut yielded the following lengths: 7.81, 7.80, 7.79, 7.80, 7.82, 7.80, 6.78, 7.80, 7.81, 7.83, 7.79, 6.77, 7.82, 7.80, 7.78, 7.80, 7.81, 7.81, 7.79, 7.87. Plot the frequency distribution plot, and determine the standard deviation. What is the maximum tolerance if you wish to assure that 95.45% of the rods are acceptable?

7.17 Bags of flour at a certain manufacturing facility are filled from a spring-loaded trapdoor that is opened for a certain time period. The bags of flour weigh, on average, 5 kg, but the standard deviation is 100 g. Considering that the plant produces 10,000 bags per day, determine the number of bags of flour weighing less than 5 kg; the number of bags of flour weighing more than 5.1 kg; and the number of bags of flour between 4.9 and 5.0 kg.

7.18 In a construction project many concrete samples are taken. For one particular project 3000 samples were taken, and they had an average compressive strength of 3500 psi. The standard deviation of the group was 250 psi. Determine the number of samples with compressive strengths less than 3000 psi and the number of samples between 3250 psi and 3750 psi.

7.19 At the end of the first semester of the freshman year an engineering department found that the students had the following age distribution.

Age (years/months)	Number of persons
18/1−3	2
18/4−6	16
18/7−9	32
18/10−12	30
19/1−3	27
19/4−6	21
19/7−9	17
19/10−12	5
20/1−3	6
20/4−6	8
20/7−9	3
20/10−12	0

Plot a histogram of the data and determine the standard deviation. Can you assume a normal distribution for these data?

7.20 Determine the equation of the straight line that best represents the following (x, y) data pairs: 3, 5; 4, 4.50; 5, 6.50; 6.5, 7.0; 8.0, 8.0; 9.0, 9.50; 11.0, 12.0; 13.0, 11.50.

7.21 The temperature along a metal pipe is given by the following temperature and length data.

Temperature (°F)	Distance (in)
180	4.3
360	7.5
540	11.4
720	15.4
900	19.9

Determine the equation of a straight line representing these data.

7.22 The following data are from measuring the force and deflection data from a spring. Determine the equation of a straight line representing the data.

Force (lbf)	Deflection (in)
10	0.1
20	0.21
30	0.25
40	0.38
50	0.49
55	0.56

7.23 An orifice in a pipe has a coefficient determined by the equation

$$C = \frac{4W}{\pi D^2 t} \sqrt{\frac{1}{2g_c \rho \Delta p}},$$

where C is the coefficient, W is the weight, D is the diameter, t is the time, ρ is the density, Δp is the pressure drop, and g_c is a constant. The following information is known about the accuracy of the measurement: the weight to 1%, the diameter to 0.2%, the time to 1%, the density to 0.2%, and the pressure to 1%. Determine the percent error in the coefficient. Which term would you invest more money in to improve the accuracy of the coefficient?

7.24 In heat transfer the Nusselt number, Nu, is given by the following equation,

$$Nu = 0.023 \left(\frac{vD}{\nu}\right)^{0.8} (Pr)^{0.4},$$

where v is the velocity, D is the diameter, ν is the kinematic viscosity, and Pr is the Prandtl number. To determine the Nusselt number the following data were obtained:

$$D = 0.03 \text{ m within } 0.001 \text{ m}$$
$$v = 50 \text{ m/s within } 3 \text{ m/s}$$
$$\nu = 1.006 \times 10^{-6}, \text{ a constant}$$
$$Pr = 7.0 \text{ within } 0.2 \text{ (dimensionless units)}$$

Find the value of the Nusselt number and the percent error in its value.

CONCEPTS OF PROBLEM SOLVING AND ENGINEERING DESIGN

C H A P T E R   O B J E C T I V E S

- To analyze a problem from its structure.

- To present results in a professional manner.

- To become familiar with the unit systems most often encountered in engineering practice.

- To learn decision analysis techniques.

- To study the design process, attuning yourself to the steps involved.

- To initiate your own preliminary designs.

(Photo courtesy of NASA.)

Your course content in engineering may be thought of as having two components, analysis and design. Analysis is another name for problem solving, design another name for creativity. In this chapter you will learn about these two fundamental aspects of your engineering education.

8.1 PROBLEM FORMAT

Engineers solve problems. We are educated to solve them and like to solve them. A great deal of your education will be spent in solving a variety of problems, and you must learn to do this well and present the results in a fashion that is readily understood by the instructor and by you. Yes, you. You will use your homework problems when reviewing for tests and as references in later courses. You must include sufficient detail to make them useful to you when the material is not fresh in your mind.

Some fairly common formats are used to present homework and quiz and test results; your instructor will describe variations that she or he prefers.

Figure 8.1 illustrates one type of solution format. You may be asked to copy the problem statement or, as in this case, write what's given and what you are to find. All these formats have general similarities.

1. Always show units, such as *lbf,* on your answers, unless the term has no units.
2. Show all your calculations clearly so you and your instructor know what you are doing.
3. When you have arrived at an answer, check its reasonableness. Is it the same order of magnitude as the other terms? Check your work, assumptions, and calculations.
4. Clearly identify your answer, underlined several times.
5. Do not start another problem on the same page unless it can be finished on that page.

Showing the units on the equations will help you in defining the problem; for instance, all the terms in a force balance should have units of force. Presenting the results in this somewhat formal fashion will require you, in many instances, to do a rough draft and then copy it over. Do not expect that the first attempt will always be presentable. The calculational procedure will help your instructor

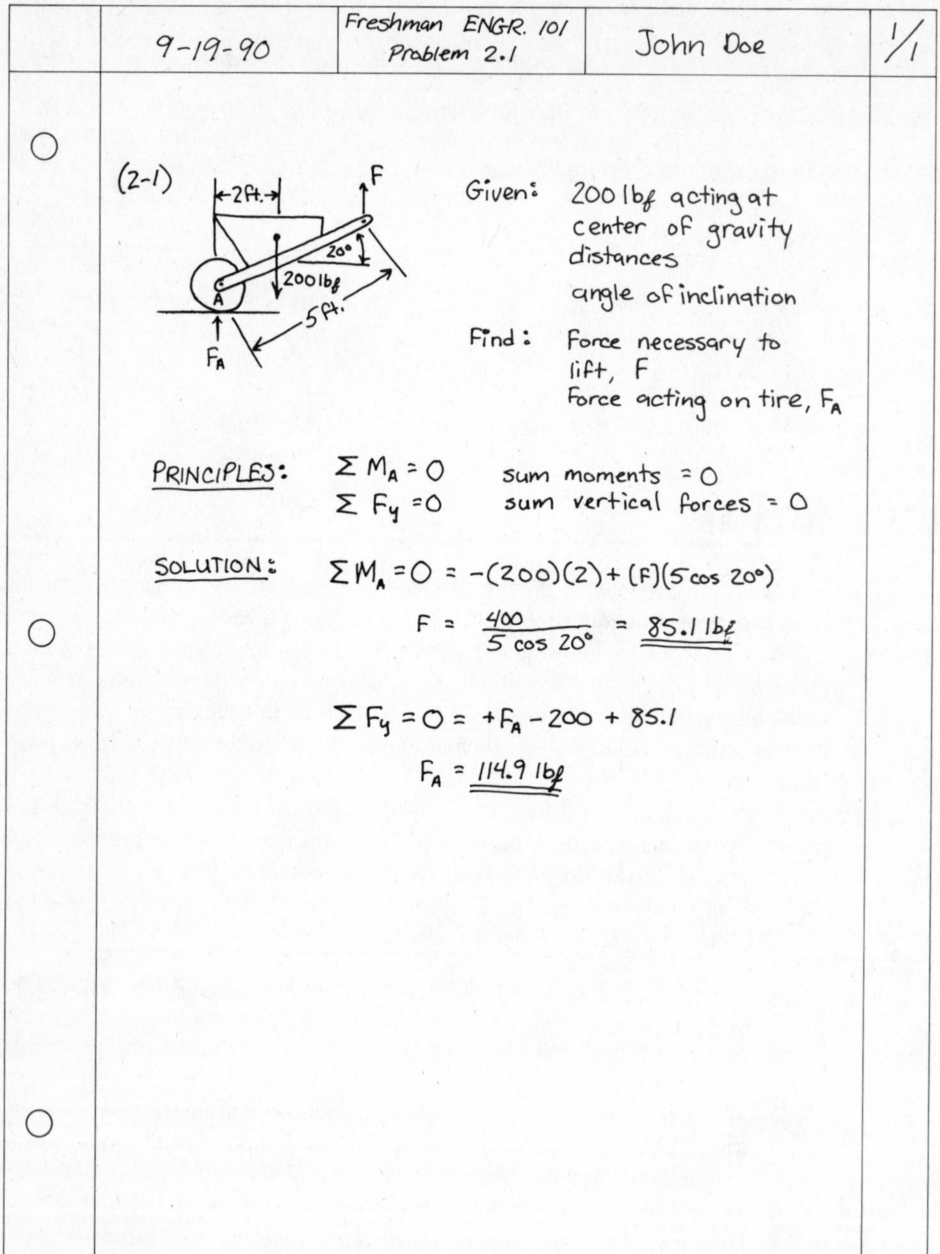

9-19-90	Freshman ENGR. 101 Problem 2.1	John Doe	1/1

(2-1)

Given: 200 lbf acting at center of gravity
distances
angle of inclination

Find: Force necessary to lift, F
Force acting on tire, F_A

PRINCIPLES: $\Sigma M_A = 0$ sum moments = 0
$\Sigma F_y = 0$ sum vertical forces = 0

SOLUTION: $\Sigma M_A = 0 = -(200)(2) + (F)(5 \cos 20°)$

$$F = \frac{400}{5 \cos 20°} = \underline{\underline{85.1 \text{ lbf}}}$$

$$\Sigma F_y = 0 = +F_A - 200 + 85.1$$

$$F_A = \underline{\underline{114.9 \text{ lbf}}}$$

FIGURE 8.1 A sample format for problem solution.

readily pinpoint a difficulty. If it is caused by a conceptual misunderstanding, you will be able to comprehend where the difficulty occurred in the context of the entire solution.

A misplaced decimal point can cause strange numbers to appear in part of the solution; see if the answer makes sense physically. Practice in problem solving will also develop your engineering intuition. It is discouraging to have numerical results that seem wrong; you end up doubting your solution procedure when indeed only a calculation error occurred. Identify the answer for all to see. Remember you will be using your homework as a study guide later, so make it worthwhile. Minimizing the number of problems on a sheet also helps when you study the material later. Problems running over to another sheet require you to constantly turn between two pages, sometimes losing your continuity of thought. Very often students are cheap when it comes to using paper, and while it is laudable that we conserve our natural resources, white areas around the sketch and numbers help tremendously in the problem's readability.

8.2 PROBLEM SOLVING

The assumption above is that you know how to solve a problem, but perhaps this is not the case. Guidelines in problem solution that have helped others are implicit in the solution shown in Figure 8.1.

1. Define the system under consideration—the material, circuit, or system that is to be analyzed: the wheelbarrow in Figure 8.1.
2. After you define the system, visualize what is happening. Drawing a sketch helps tremendously with the visualization process. In Figure 8.1 you are lifting the load, sharing a portion with the wheel.
3. Identify the given (known) material and the unknowns to be solved for.
4. Model the system with appropriate equations, such as the sum of the forces must be zero for equilibrium.
5. Perform the analysis necessary to solve for the unknowns (not necessarily an easy task).

Problem solving is not easily learned just by following steps. The ones mentioned above give us a direction to proceed in, but the path in that direction is not always clear. We can increase the clarity if we understand some of the conceptualizations that must occur before we can intelligently, rather than by rote, solve a variety of problems. Much of what you have learned about problem solution was developed in your high school algebra classes. Having a sound algebraic ability will help you in engineering.

The intuitive technique for problem solving has four steps that often happen so quickly we don't realize they have occurred. Simply stated they are

1. Understanding the statement of the problem (a sketch helps),
2. Finding what is required,
3. Finding what facts are given, and
4. Deciding what operations are needed.

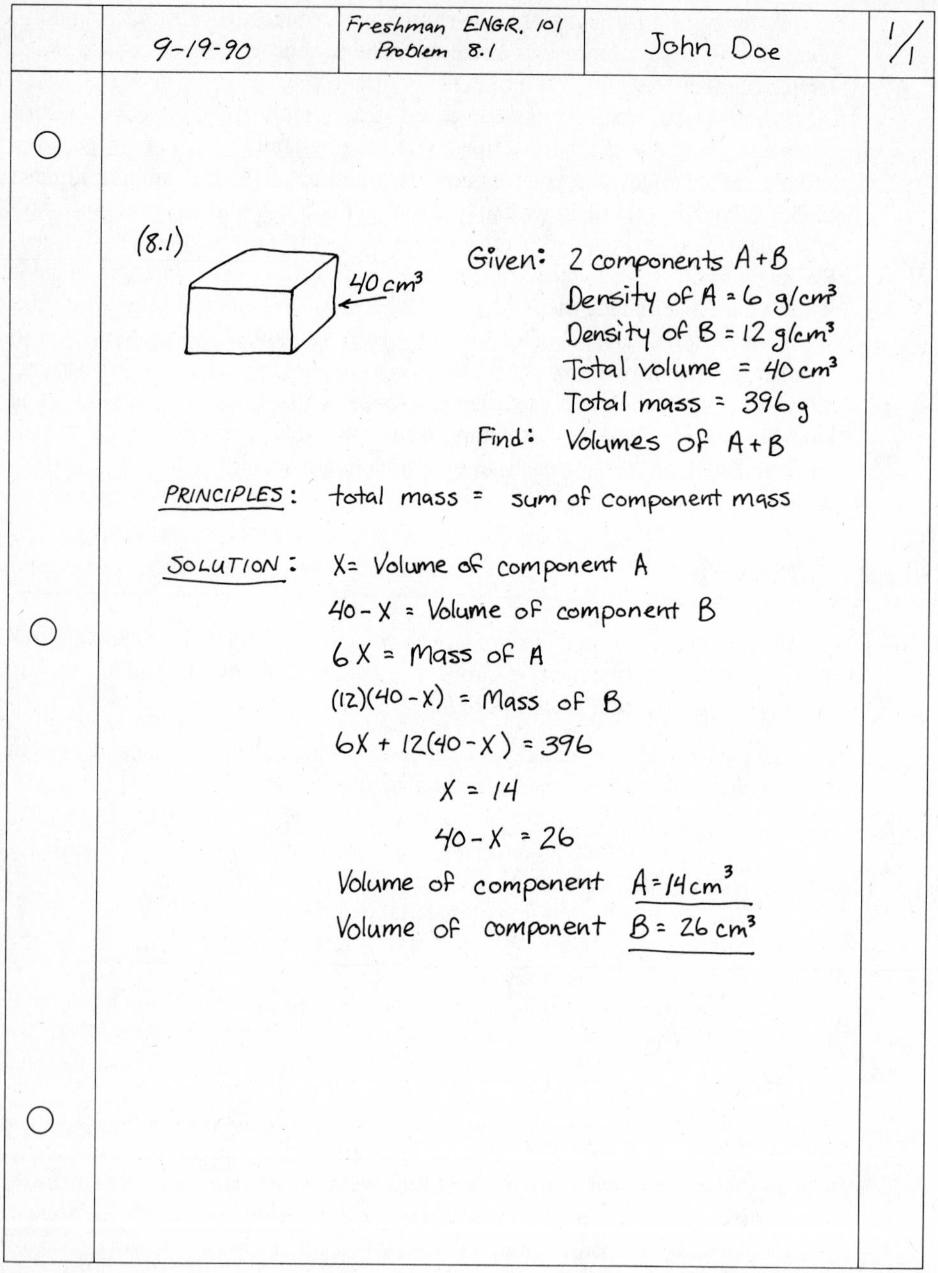

(8.1)

40 cm³

Given: 2 components $A + B$
Density of $A = 6$ g/cm³
Density of $B = 12$ g/cm³
Total volume $= 40$ cm³
Total mass $= 396$ g

Find: Volumes of $A + B$

PRINCIPLES: total mass = sum of component mass

SOLUTION: $X =$ Volume of component A

$40 - X =$ Volume of component B

$6X =$ Mass of A

$(12)(40-X) =$ Mass of B

$6X + 12(40-X) = 396$

$X = 14$

$40 - X = 26$

Volume of component $\underline{A = 14 \text{cm}^3}$
Volume of component $\underline{B = 26 \text{cm}^3}$

FIGURE 8.2 The solution to Example 8.1.

This is a more general rendition of the steps outlined previously. This listing stresses two important aspects that you might have missed earlier. From the first step you must understand the language that is used to describe the problem; you must be able to visualize (sketch) it if it is a physically real system. Part of your course material will help you to understand the language and its implications, and you will be able to visualize systems correctly. If you cannot, see your instructor. You will have pinpointed an area of confusion, so he or she can more easily be of assistance. You will not be successful in problem solving if this visualization and understanding are lacking.

Normally in textbook problems the information given (Step 3) and what is required (Step 2) are clearly identified. In some advanced areas this is not the case. Step 4, however, often lacks clarity. How do you decide what operations are required? Poor decisions here are most often the cause for failure. Once the equations are written, you and others can usually solve them.

Certain techniques will enable you to make better decisions. The key word is *connections,* the various connections between the different quantities in the problem statement. Let's see what this means by looking at an example.

**E X A M P L E
8.1**

The volume of a solid is 40 cubic centimeters. The solid is composed of two substances, one with a density of 6 g/cm^3 and the other with a density of 12 g/cm^3. The solid weighs 396 grams. Determine the volume that each substance occupies.

SOLUTION:

The first reading of the problem should bring into focus the differences among quantity, number, and units. Density is a quantity, its numerical value is 6 for one substance and 12 for the other, and it has units of g/cm^3. Thus, we must understand philosophically what density is, what the units of it are in this particular situation, and what numerical value is associated with those units. With the first reading we look to establish a connection between the quantities that are given in the problem statement, Step 1. In this case there are four quantities involved: components in volume, mass per unit volume of component, total volume, and total mass.

Next we establish the equations that relate to the connections between the quantities. If we multiply the component density by the component volume, we obtain the component mass. This, when added to the other component's mass, will give the total mass. Now we can proceed to solve the problem according to the general guidelines given above. Figure 8.2 illustrates the solution.

These reasoning techniques enable you to take the information given in the problem statement and manipulate it. In some instances reasoning, or processing the information, is not the difficulty; gathering the information creates the trouble. Figure 8.3 illustrates an information diagram of two general skills, processing information (reasoning) and gathering information. Reasoning ability involves inductive and deductive thinking and logic. Information gathering ability relates to reading comprehension, visualization, memory, and recall.

We would all like to be in quadrant I, have good abilities in reasoning and information gathering. However, in some courses you may find that you possess such abilities while in others you do not. The instructor may affect your attitude, hence ability, about the subject matter, making it easier or more difficult to assimilate. In other instances you need to pay special attention to developing your ability in either the gathering or processing areas.

Quadrant II denotes people who gather the information correctly, take good notes, understand what they read, and know what is asked of them, but have difficulty knowing how to proceed with the problem solution. For these people, following the formal problem solving procedures will be of assistance. Checking with the instructor about the logic of the solution is clearly important.

People in quadrant IV have the opposite problem. They can process the information correctly, once they know what the information is. Skills in note taking, reading comprehension, memorization, and visualization of what is happening are usually important in strengthening this skill. These people often need to focus on understanding technical words and the implications of certain terms.

People in quadrant III have to work hard to overcome deficiencies in two skill levels. Like those in quadrant IV, they must improve their information gathering skills, and then learn to process this information correctly, like those in quadrant II.

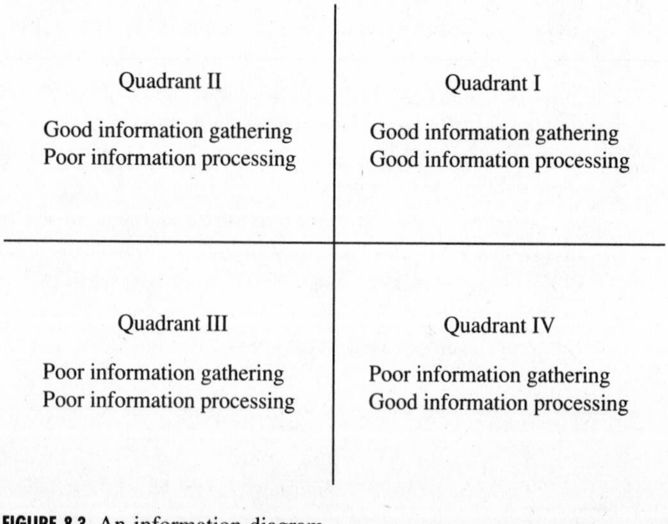

Quadrant II	Quadrant I
Good information gathering Poor information processing	Good information gathering Good information processing
Quadrant III	Quadrant IV
Poor information gathering Poor information processing	Poor information gathering Good information processing

FIGURE 8.3 An information diagram.

Both aspects of problem solution can be improved by being aware of what it is you do not understand. Is it the vocabulary? The visualization? Or does the difficulty lie in what to do with the information once you have it? By answering these questions you can focus your efforts on the area of weakness.

8.3 SIGNIFICANT FIGURES AND SCIENTIFIC NOTATION

The use of scientific notation for numbers is a great help when using large numerical values. For instance, if the energy used in a power plant is 185 million kilowatts and this number has to be used in calculations, it becomes cumbersome to write and manipulate. Expressed in powers of ten (scientific notation—see Table 8.1), the number is easier to work with: 1.85×10^8.

To use scientific notation accurately and correctly we need to understand the concept of significant figures. Suppose the result of a calculation is 7.68321, using a hand calculator. Are all those numbers after the decimal important, useless, or misleading? The numbers used in a calculation came from somewhere, resulted from measurements of some kind. The accuracy of the measurement determines the number of significant figures, the degree of correctness of the number. If you are measuring the thickness of an electric cable, you might use a ruler and find it is three-quarters of an inch in diameter. How sure are you of this measurement? As accurately as you can read the ruler, in this case 0.75 inches within a thirty-second of an inch, or ±0.03 inches. Thus, you would give 0.75, as that is the degree to which you are certain of the measurement. It would be erroneous to write 0.750, as that implies you know the value to the thousandth of an inch. Or you might use a micrometer and determine the measurement to be 0.740 ±0.002; in this case your reading is accurate to the

TABLE 8.1 SCIENTIFIC NOTATION

1 000 000	=	1×10^6
100 000	=	1×10^5
10 000	=	1×10^4
1 000	=	1×10^3
100	=	1×10^2
10	=	1×10^1
1	=	1×10^0
0.1	=	1×10^{-1}
0.01	=	1×10^{-2}
0.001	=	1×10^{-3}
0.0001	=	1×10^{-4}
0.00001	=	1×10^{-5}
0.000001	=	1×10^{-6}

nearest thousandth of an inch, and the extra decimal place should be used to convey this information.

Number	Significant figures
832	3
19	2
21.61	4
0.621	3
401.612	6
0.02	1
0.038	2
0.02000	4

Returning to scientific notation, the steps involved for expressing a number are

1. Locate the decimal point so the number has a value between one and ten.
2. Use only the number of digits that are significant figures.
3. Multiply the number by powers of ten to obtain the correct magnitude.

The value of 986 in scientific notation is 9.86×10^2; the value of 986.00 is 9.8600×10^2 because the decimal point in the original figure indicated the greater accuracy of the number, five significant figures versus three for 986. If the number is given as 98,600, we cannot tell whether the number should be written as 9.86×10^4, 9.860×10^4, or 9.8600×10^4. The initial number does not convey its significant figures accurately. In textbooks, an immediate concern of yours, consider all numbers to have the maximum significant figures possible unless you are given information to the contrary.

The value of scientific notation is that it allows you to inform others as to the accuracy of the number, its significant figures. In addition it allows you to handily express large numbers used in calculations. Hence, $5,630,000 \times 0.00000609 = 34.3$ becomes $(5.63 \times 10^6) \times (6.09 \times 10^{-6}) = 3.43 \times 10^1$.

For addition, subtraction, multiplication, and division of numbers with varying significant figures, the result is only as accurate—has as many significant figures—as the number with the least significant figures. Thus, when we add the following figures, $28.631 + 4928.1 + 0.9763 = 4957.7073$, the second term is accurate to only the tenths place, so the correct answer is 4957.7. Multiplication and division follow the same rule. Thus, a calculator might display the following result, $27.3 \times 0.93651 = 25.566723$, but the corrected answer is 25.6. The other numbers are rounded off.

The rule for rounding off is that if the digit to the right of the last significant figure is five or greater, round the last significant figure up; if it is less than five, round the figure down. If 31.6952 and 34.6949 are rounded off to four significant figures, the results are 31.70 and 34.69. Conversion factors (2.54 centimeters per inch) or integer values used when performing arithmetic operations are considered to contain unlimited significant figures. The other terms dictate the number of significant figures in the answer.

8.4 UNIT SYSTEMS

The numbers engineers work with in most instances have dimensions associated with them. We use these numbers to characterize systems by their length, mass, temperature, and other standards. When these standards are combined into a coherent set, they form a measurement system. The English engineering system and the SI metric system are the two systems most frequently encountered in engineering practice and in college texts. Other systems are used; however, if you understand the bases for these two, others are similar in terms of their development.

Fundamental and Derived Units

Units that are postulated, or defined, are called fundamental units. Once a few units are defined, others can be derived from them. Some of the fundamental units are length, time, mass, force (in the English system), and temperature.

Length (L) is the distance between two points in space.

Time (t) is the period between two events or during which something happens.

Mass (m) is the quantity of matter that a substance is composed of. It is invariant with location: the mass of a person on earth or on the moon is the same.

Force (F) is defined through Newton's second law; conceptually it is associated with power or strength, the push or pull on an object. It may occur directly, such as the push on a wheelbarrow, or indirectly by fields (gravitational, electrical, magnetic) acting on an object. Force is sometimes viewed as a fundamental unit, as it is in the English engineering system, and sometimes as a derived unit, as in the SI system.

Temperature (T) measures the degree of hotness or coldness of an object. We commonly use arbitrary temperature scales, such as the Fahrenheit and Celsius scales. Temperature scales are for the most part based on two points, the triple point of water (where ice, liquid, and vapor coexist) and the boiling point of water at atmospheric pressure. The Fahrenheit scale was named for Gabriel Daniel Fahrenheit. He was interested in thermometry, as was an astronomer friend, Romer, who devised a temperature scale of 60 degrees. There are 60 seconds to a minute, 60 minutes in an hour, so why not 60 degrees for a temperature scale? Fahrenheit thought the scale too rough, and increased the number of divisions to 240; why not 360 divisions remains a mystery. The lower limit was an ice, salt, and water mixture that corresponds to 0°F, and the human body temperature was set at 90°, which Fahrenheit later decided should be 96°. With the scale fixed by body temperature and an ice, salt, and water mixture, the triple point became 32° and water boiled at 212°. To eliminate these inconvenient numbers a Swedish astronomer by the name of Anders Celsius devised a scale that started at 100° (triple point) and went to 0° (boiling point). A friend suggested he reverse them, which resulted in the current Celsius temperature scale. Lord Kelvin developed the idea, championed by others earlier, of the absolute temperature scale and devised the Kelvin temperature scale, which uses the divisions of the Celsius scale. Another absolute temperature scale is the Rankine scale, which uses the divisions of the Fahrenheit scale. The relationships between the scales are

$$F = 9/5 \ C + 32$$
$$C = 5/9 \ (F - 32)$$
$$K = C + 273.16$$
$$R = F + 459.67$$

where F is degrees Fahrenheit, C degrees Celsius, K degrees Kelvin, and R degrees Rankine. Figure 8.4 graphs the four temperature scales.

Other units can be derived from fundamental units. Area and volume are derived from length, L^2 and L^3. Velocity is distance per unit time, or L/t, and pressure is force per unit area, or F/L^2. Unit systems assign numerical values to these dimensions, and we normally use these dimensions in conjunction with the unit systems, merging the two concepts in our minds.

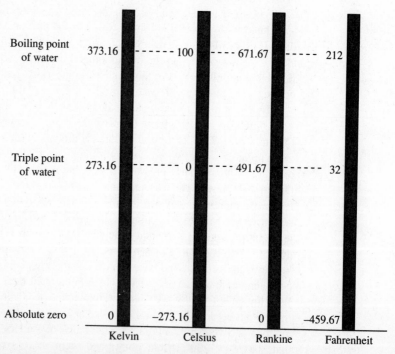

FIGURE 8.4 The graph of four temperature scales.

English Engineering Unit System

The English engineering unit system developed in a rather haphazard fashion, with many common measures, such as the yard, descended from such standards as the length from the thumb to the nose of Henry I of England. Clearly, such a system has many drawbacks, but a singular advantage is that we know it well, having grown up with feet and pounds.

One drawback is the confusion that results from having force and mass as fundamental units. They are related through Newton's second law, which states that the force acting on a body is equal to the mass times acceleration.

$$F = ma, \qquad\qquad 8.1$$

where F is measured in pounds force (lbf), m is measured in pounds mass (lbm), and a is measured in feet per second squared (ft/sec^2). One pound force is defined as one pound mass accelerated 32.174 ft/sec^2. When we substitute the values of the definition into Equation 8.1, a problem appears in the unit balance of the equation: lbf = (lbm)(ft/sec^2). The units on one side do not equal the units on the other side, so let's multiply Equation 8.1 by a constant that has units that balance the equation:

$$F = kma, \qquad\qquad 8.2$$

or with the units substituted,

$$lbf = k(lbm)(ft/sec^2).$$

Thus, the units of the constant must be lbf-sec^2/lbm-ft. To determine the value of the constant, we substitute into Equation 8.2 the definition of one pound force and solve for k:

$$k = \frac{1}{32.174} \times \frac{lbf\text{-}sec^2}{lbm\text{-}ft}. \qquad\qquad 8.3$$

Since Equation 8.2 is valid for all cases, if k is determined for a given case, it must be true for all the other cases. Standard gravitational acceleration, g, is 32.174 ft/sec^2, a term with the same numeric value as k but different units. The term k was noted to be equal to $1/g_c$, where $g_c = 32.174$ lbm-ft/lbf-sec^2, and thus Equation 8.2 becomes

$$F = ma/g_c \text{ lbf.} \qquad\qquad 8.4$$

■

E X A M P L E
8.2

Consider that a man seated in a car is uniformly accelerated at 5 ft/sec^2 and that the man has a mass of 180 lbm. Find the horizontal force acting on the man.

SOLUTION:
$$F = ma/g_c = \frac{(180 \text{ lbm})(5 \text{ ft/sec}^2)}{32.174(\text{lbm-ft})/(\text{lbf-sec}^2)}$$

■

TABLE 8.2 UNITS IN THE ENGLISH ENGINEERING SYSTEM

Fundamental		Derived	
Force (F)	lbf	Area (L^2)	ft^2
Mass (M)	lbm	Volume (L^3)	ft^3
Length (L)	ft	Acceleration (L/t^2)	ft/sec^2
Time (t)	sec	Density (M/L^3)	lbm/ft^3
		Pressure (F/L^2)	lbf/ft^2
		Energy (FL)	ft-lbf
		Power (FL/t)	(ft-lbf)/sec

Thus far we have not mentioned weight. Take the man in example 8.2 and let him be standing on the ground. He experiences a force on him equal and opposite to the force he exerts on the earth. This force is his weight. He is acted upon by the local gravitational field g. If the gravitational field is located so g has a value of 32.174 ft/sec^2, the standard gravitational acceleration, then the force is

$$F = \frac{(180 \text{ lbm})(32.174 \text{ ft/sec}^2)}{32.174 \text{ (lbm-ft)/(lbf-sec}^2)} = 180 \text{ lbf}.$$

If the man were standing on a tall mountain where g has a value of 30.0 ft/sec^2, then the force exerted on the ground, his weight, would be

$$F = \frac{(180 \text{ lbm})(30.0 \text{ ft/sec}^2)}{32.174 \text{ (lbm-ft)/(lbf-sec}^2)} = 167.8 \text{ lbf}.$$

In space, the condition of weightlessness is caused by the fact that $g = 0$; hence the force or weight of the body mass is zero.

Table 8.2 lists the fundamental and some of the derived units found in the English engineering system.

SI Units

SI stands for Système International d'Unités, a consistent unit system developed in 1960. In 1975 the metric conversion act became law, declaring in part "a national policy of coordinating the increasing use of the metric system in the United States." The process of conversion has been slow, but as the business outlook of the United States becomes more global, the United States will convert of necessity to metric sizes. It is easier to perform calculations in metric than in English engineering units. The difficulty in the conversion process is that standard sizes of equipment, pipes, bolts, and parts of all kinds must be changed. This requires a major investment in new manufacturing tools, which businesses may not be able to afford, but eventually they must convert if they are to compete in the world marketplace where metric is standard.

TABLE 8.3   FUNDAMENTAL UNITS IN SI

Quantity	Name	Symbol
Base units		
Length	meter	m
Mass	kilogram	kg
Time	second	s
Electric current	ampere	A
Thermodynamic temperature	kelvin	K
Amount of substance	mole	mol
Luminous intensity	candela	cd
Supplementary units		
Plane angle	radian	rad
Solid angle	steradian	sr

SI has seven physically defined units—fundamental or base units—and two geometrically defined supplementary ones, and the rest are derived. The fundamental and supplementary units are listed in Table 8.3. Each of these terms is rigorously defined.

A meter is the distance in space that is equal to $1.650\ 763\ 73 \times 10^6$ wavelengths in vacuum of the radiation corresponding to the transition between two energy levels of the krypton 86 atom. This is not a handy definition, but is certainly a precise one. For everyday conceptualization a meter is 39.37 inches.

A kilogram is a mass equal to the mass of an international prototype made of a platinum-iridium alloy and kept at the International Bureau of Weights and Measures in France. Relating this to our everyday life, it is equal to 2.205 pounds mass.

A second is defined as the duration of $9.192\ 631\ 770 \times 10^9$ periods of radiation in the transition between the two hyperfine levels of the ground state of the cesium 133 atom. Originally a second was based on a fraction of the mean solar day, which used the earth's rotation as a clock, and this is the basis of our day-to-day observation of time.

An ampere is the constant current that, if maintained in two straight parallel conductors of infinite length and of negligible circular cross-section and placed one meter apart in vacuum, would produce between these conductors a force equal to 2×10^{-7} newtons per meter of length. Again in daily life this is essentially a flow of one coulomb per second, where one coulomb is equal to 3×10^9 units of charge.

A kelvin degree is 1/273.16 of the thermodynamic temperature of the triple point of water.

A mole is the amount of substance in a system that contains as many elemental entities as there are atoms in 0.012 kilograms of carbon 12.

A candela is the luminous intensity, in the perpendicular direction, of a surface of 1/600 000 square meter of a blackbody at the temperature of freezing platinum under a pressure of 101 325 newtons per square meter.

A radian is a unit of measure of a plane angle with its vertex at the center of a circle and subtended by an arc equal in length to the radius.

TABLE 8.4 SI UNIT PREFIXES

Factor by which unit is multiplied	Prefix	Symbol
10^{-18}	atto	a
10^{-15}	femto	f
10^{-12}	pico	p
10^{-9}	nano	n
10^{-6}	micro	μ
10^{-3}	milli	m
10^{-2}	centi[a]	c
10^{-1}	deci[a]	d
10^{1}	deka[a]	da
10^{2}	hecto[a]	h
10^{3}	kilo	k
10^{6}	mega	M
10^{9}	giga	G
10^{12}	tera	T
10^{15}	peta	P
10^{18}	exa	E

[a]Avoid these prefixes, except for centimeter.

A steradian is a unit of measure of a solid angle with its vertex at the center of a sphere and enclosing an area of the spherical surface equal to that of a square with sides equal in length to the radius.

Force is a derived unit in SI; one unit of force is defined by Newton's second law, Equation 8.1. A unit force is a newton (N), so $1 \text{ N} = (1 \text{ kg})(1 \text{ m/s}^2)$. The potential for confusion between mass and force does not exist, as they not only have different names but different numerical values as well. For instance, the force exerted on a 40 kilogram woman by standard gravitational acceleration, 9.8 m/s^2, is

$$F = (40 \text{ kg})(9.8 \text{ m/s}^2) = 392 \text{ N}.$$

Table 8.4 shows the prefixes that indicate the order of magnitude of the term and derived units in SI. Table 8.5 lists the derived units in SI.

SI symbols are strictly interpreted. SI associates an exact meaning to each symbol, and one of its values lies in this preciseness. On the other hand, it does not allow for variations from these rules, as these variations would imply some other meaning. For instance in English units we may write ft-lbf or lbf-ft and have the same meaning, though the first expression is desirable. In SI N m and mN are entirely different: the former is newton meter and the latter is millinewton. Notice that a space was left between N and m in the first term; this is the correct way to write this combination of SI units. The following is a list of some important rules for using SI.

1. Unit symbols are printed in lowercase roman letters. Periods are not used after a symbol except at the end of a sentence. Thus, we would write N

TABLE 8.5 SI DERIVED UNITS

Quantity	Unit	SI Symbol	Formula
Acceleration	meter per second squared	—	m/s^2
Angular acceleration	radian per second squared	—	rad/s^2
Angular velocity	radian per second	—	rad/s
Area	square meter	—	m^2
Density	kilogram per cubic meter	—	kg/m^3
Electric capacitance	farad	F	$A{\cdot}s/V$
Electrical conductance	siemens	S	A/V
Electric field strength	volt per meter	—	V/m
Electric inductance	henry	H	$V{\cdot}s/A$
Electric potential difference	volt	V	W/A
Electric resistance	ohm	Ω	V/A
Energy	joule	J	$N{\cdot}m$
Entropy	joule per kelvin	—	J/K
Force	newton	N	$kg{\cdot}m/s^2$
Frequency	hertz	Hz	$1/s$
Illuminance	lux	lx	lm/m^2
Luminance	candela per square meter	—	cd/m^2
Luminous flux	lumen	lm	$cd{\cdot}sr$
Magnetic field strength	ampere per meter	—	A/m
Magnetic flux	weber	Wb	$V{\cdot}s$
Magnetic flux density	tesla	T	Wb/m^2
Power	watt	W	J/s
Pressure	pascal	Pa	N/m^2
Quantity of electricity	coulomb	C	$A{\cdot}s$
Quantity of heat	joule	J	$N{\cdot}m$
Radiant intensity	watt per steradian	—	W/sr
Specific heat	joule per kilogram-kelvin	—	$J/kg{\cdot}K$
Stress	pascal	Pa	N/m^2
Thermal conductivity	watt per meter-kelvin	—	$W/m{\cdot}K$
Velocity	meter per second	—	m/s
Viscosity, dynamic	pascal-second	—	$Pa{\cdot}s$
Viscosity, kinematic	square meter per second	—	m^2/s
Volume	cubic meter	—	m^3
Work	joule	J	$N{\cdot}m$

or kg, and not N. or KG. Italic letters are used for quantity symbols; thus, m is mass and m is meter. Additionally the symbol is the same in the singular and plural: thus, 1 kg, 10 kg.

2. The prefix precedes the symbol in roman type without a space, and double prefixes are not allowed: Thus, mA and not μmA.

3. For distance in engineering and architectural drawings, use millimeter as the basic unit.

4. The product of two or more units is represented by a space or a dot between the units; thus, N m or N·m.

5. Where the division of terms must occur the slash mark should not be used more than once, nor should a hyphen be repeated on the same line, as ambiguity results in both cases. In the case of acceleration, m/s² or m-s⁻² are acceptable, but m/s/s or m-s-s is vague. It could be interpreted that the seconds cancel one another and meters remain.

6. Capital and lowercase letters must be used as described; for instance, K means degrees Kelvin and k means kilo.

7. Don't use commas in long numbers. In some countries commas are used in place of decimal points. Rather, use a space after every third digit to the left and right of the decimal point. If there are only four digits the use of the space is optional. Thus, you would write 3 600 000 and 21.005 68, not 3,600,000 or 21.00568.

8. When writing a number less than one, always put a zero before the decimal point; 0.625, not .625.

Some of these rules are difficult to use when you are writing by hand, particularly differentiating between roman and italic letters; adopt a different symbol for clarity, such as a script *m* for mass, so your notes are not confusing.

8.5 CONVERSION FACTORS

It will remain necessary to convert between unit systems for many years, as different unit systems are in use around the world. A conversion factor helps us convert from the value in one system to its value in another. Appendix 2 gives a list of conversion factors, and we will shortly demonstrate their use in several situations. But more fundamentally we should understand why they can be used: they are dimensionless ones, and hence we can multiply or divide any equation by them and not change the value of the equation. Let's examine the conversion between inches and centimeters, 1 inch = 2.54 centimeters. Divide this equation by 1 inch, yielding 1 = 2.54 cm/in. Thus, 2.54 cm/in is equal to a dimensionless one.

■
E X A M P L E
8.3

A cargo ship has large tanks within it for carrying fuel oil. The oil is measured in barrels, while the tank dimensions are in meters, 1 m × 5 m × 15 m. How many barrels does the tank hold? How many gallons?

SOLUTION:

Calculate the volume in cubic meters; then find in the conversion tables the value to convert from cubic meters to barrels.

$$V = 1 \times 5 \times 15 = 75 \text{ m}^3$$

The direct conversion is not listed, so we must convert from cubic meters to gallons and from gallons to barrels.

$$V = 75(2.642 \times 10^2) = 19,815 \text{ gallons}$$

Again we must calculate a new conversion factor, as we are given that 42 gallons equals 1 barrel. The conversion factor is 1/42 barrels/gallon, and the volume in barrels is

$$V = (19{,}815)/(42) = 471.8 \text{ barrels.}$$

■

8.6 TECHNICAL DECISION ANALYSIS

Engineers at all levels, from entry level to senior management, need to make decisions, assessing the various factors underlying the decision and then reaching a logical conclusion about it. As an engineering manager you might have to decide whether or not to invest in a new product line with its related machinery costs, or use the money elsewhere. There will be no clear-cut yes-or-no answer: uncertainty is attached to whatever choice you make. Techniques that assist in decision making will be investigated here. As a manager you will be judged by the decisions you make and, unlike a baseball batting average, judged by the last few decisions and whether or not they were correct, not by the total number of correct decisions. You must make decisions in situations where you will have insufficient information or imperfect information. You no longer need to shoot from the hip, however, as methodologies exist to minimize risk to any decision and maximize return.

Decision Tree

Your company owns land on which there may or may not be oil. A competitor of yours is willing to lease the land from you for $500,000, and you must make a decision within ten days. If you accept the lease, the company receives $500,000. If you reject the proposal, you have two choices, to drill for oil or not. If you decide to drill for oil there are four possible outcomes: natural gas, no oil, occurring 40 percent of the time; natural gas and oil, occurring 30 percent of the time; dry hole, occurring 20 percent of the time; and oil, occurring 10 percent of the time. The cost of drilling is $1,500,000, and from past experience you know that the value gained by the company is zero for the dry hole; $1,500,000 for natural gas only; $3,000,000 for gas and oil mixture; and $6,000,000 for oil only. What decision should you make, assuming that incurring a loss of $1,500,000 will not bankrupt the company?

To help answer this question you might draw a decision tree, as shown in Figure 8.5. This tree represents all the possible decisions, their outcomes, and the probability of each outcome. Initially there are two choices, to accept the lease or not to do so. If the lease is rejected, again there are two choices, to drill or not to drill. If you drill there are four possible outcomes; the net dollar value is listed at the far righthand side of the tree. How then to decide? The various

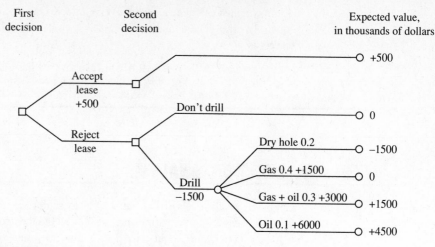

FIGURE 8.5 A decision tree.

options in the ''to drill'' branch may be combined into an average expectation by finding the weighted probability and the expected value (EV) associated with it.

$$EV = 0.2(0 - 1{,}500{,}000) + 0.4(1{,}500{,}000 - 1{,}500{,}000)$$
$$+ 0.3(3{,}000{,}000 - 1{,}500{,}000) + 0.1(6{,}000{,}000 - 1{,}500{,}000)$$
$$EV = +600{,}000$$

Thus, the expected value from drilling, considering all the various possibilities and their probabilities, is $600,000. You would choose then not to lease. The simplified decision tree is shown in Figure 8.6. You could still come up with a dry hole and lose $1,500,000, but perhaps there is a way to minimize this possibility by obtaining more information about the land by test drilling. This information has a cost, and it makes for a much more complicated decision tree,

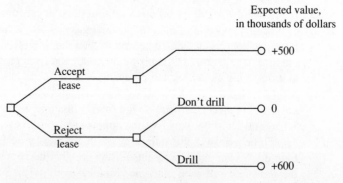

FIGURE 8.6 A simplified version of the decision tree in Figure 8.5.

but the value of the information can be quantified to some degree by this analysis.

The area of decision analysis is very interesting and useful in estimating the value of a given decision. Although the example is oversimplified, it should give you an idea of what is involved. Industrial engineering, operations research, and business administration deal with problems of this type.

Multicriteria Decision Analysis

The previous example had one criterion for judging the merits of the various courses of action, the expected value. Often a situation has several criteria, of various degrees of importance, which must be evaluated simultaneously. This requires multicriteria decision analysis.

Consider the situation where you have job offers from two totally different companies. How might you decide which position to accept? First establish the criteria you will use in making a decision. Then determine the relative importance, the weighting factor, associated with each one. Table 8.6 lists the criteria and their weights. Determining the criteria and the weights is, in general, not an easy task if skewing the decision is to be avoided. For this example five criteria have been selected: salary and benefits are very important to you and are given a weight of nine; location of the company—what part of the country, nearness to your family—is also important and is given a weight of seven; the cost of living affects how much of your income you will be left with after the essentials are taken care of and is given a weight of five; important to your career development is the opportunity for advancement within the organization and is given a weight of seven; certainly how you relate to your co-workers effects your decision in selecting the company, hence the weighting of five.

You receive job offers from companies A and B and the salary for A is 15 percent better than that offered by B. The location is not as acceptable as B, though not disastrous; the cost of living is about the same in both locations; the route of advancement at A is more regimented than at B; and you liked the people you would be working with at B slightly more than at A. Table 8.7 shows the final analysis in rating the companies. You determine the relative grade each

TABLE 8.6 CRITERIA AND WEIGHTING FACTORS IN JOB SELECTION

Criteria	Weight (1–10)
1) Salary and benefits	9
2) Location	7
3) Cost of living	5
4) Opportunity for advancement	7
5) Co-worker compatibility	5

TABLE 8.7 RESULTS OF MULTICRITERIA EVALUATION

Company A (1–10)			Company B (1–10)		
1)	$10 \times 9 =$	90		$8 \times 9 =$	72
2)	$5 \times 7 =$	35		$7 \times 7 =$	49
3)	$6 \times 5 =$	30		$6 \times 5 =$	30
4)	$6 \times 7 =$	42		$8 \times 7 =$	56
5)	$7 \times 5 =$	35		$8 \times 5 =$	40
Total		232			247

company received for each criterion, multiply by the weighting factor, and total. In this case company B is the choice.

Gantt Chart

Managing a project with interdependent operations, such as the construction of a building, requires the scheduling and controlling of various trades and material flows. From such a schedule, situations that may cause bottlenecks can be ascertained and extra supervision provided at these junctions. The simplest form of schedule is the Gantt chart, shown in Figure 8.7. As work is completed you color in the areas, allowing you to see the timeliness of the project. In Figure 8.7 the work is slightly ahead of schedule.

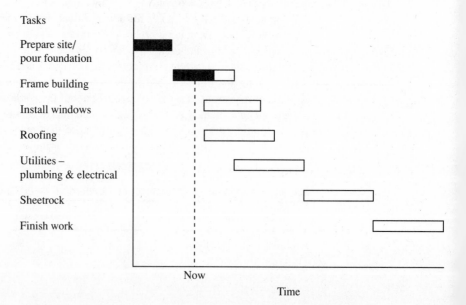

FIGURE 8.7 A Gantt chart of the schedule for the construction of a building.

This method is useful for small projects but would become too cumbersome for those with many tasks. Nor does it indicate which parts of the schedule can fall behind and not hinder the progress of the project and which parts are critical. The methods that take these factors into consideration, such as critical path analysis, are beyond the scope of this text.

8.7 ENGINEERING DESIGN

In Chapter 3 one of the various careers in engineering that we investigated was design engineering. You will find that design forms a vital part of your education in engineering, no matter what your special field. The engineer transforms technical knowledge through inspiration, perspiration, and creativity into new products and systems that satisfy a need, solve a problem. This is engineering design. All fields of engineering require engineers to display this ability: constructing new buildings and roadways in civil engineering; building advanced rockets and aircraft in aerospace engineering; and devising new microchips and computer systems in electrical engineering.

When you seek and obtain your first job, companies do not expect you to know the leading edge of development in a given field. They can educate you quickly to this level. They cannot, however, teach design and rely on your educational program to develop your ability in this area. Hence, you need the design courses and design aspects in many of the courses you take as juniors and seniors.

8.8 THE DESIGN PROCESS

The design process is composed of several steps: stating the problem, creating possible solutions, analyzing them, evaluating them, and implementing the design solution.

1. Problem definition
2. Creating solutions
3. Analysis of solutions
4. Solution evaluation
5. Problem resolution/Solution implementation

Problem Definition

In textbooks the problems are usually nicely defined, that is, given this, find that. In the engineering world, however, such clean definition is not always possible, and if the definition is determined, finding what is known or what can be

Problem definition was no easy task when considering landing an astronaut on the moon's surface. A variety of factors, some of which could only be guessed at, had to be considered before solutions were created. (Courtesy of NASA)

assumed is often perplexing. On large projects there can be underlying questions concerning time constraints on implementation; some solutions won't work because they are too time intensive. When dealing with a subsystem or a component within a total system, defining the boundaries is of critical importance. There are compatibility problems to consider; specifications must state the input and output requirements of the unit. These become the criteria by which the design is judged in later steps.

Consider that you are called upon to design the coal loading system for railroad cars. One underlying constraint affecting the design could be that the train must keep moving at a certain velocity during the loading process. Problems have constraints on them, parameters within which solutions must function.

Creating Solutions

The next phase of the design process is creating solutions for the problem. Notice that solutions, the plural, is used; there will be several ways to resolve the problem, some better than others, but they all should be examined initially. Do not prejudge or censor ideas before allowing them to surface. This is where your technical education and engineering experience combines with your innate creative ability to develop problem solutions. The ways that have been used in the past may be just fine for the present situation: no new technology or manufacturing methods are needed. However, inventiveness is required when research and development have created new technologies that can be used for new products and projects. For instance, the advances in current aircraft design could not have occurred without first developing the graphite epoxy materials that withstand greater forces than does high-strength aluminum, the material previously in use.

Often in the design process when solutions are first created, questions arise about the problem statement whose resolution better define the statement. The solutions remain qualitative at this point.

Analysis of Solutions

The analysis of solutions is another area served by your engineering education; you become proficient at mathematically modeling the various physical situations. The analysis of the model will include use of the constraints that have been developed in the problem definition. The forces acting on a structure, the electrical signal that must be interpreted, are examples of these types of constraint.

For large projects, or problems requiring advances in technology, the engineer may request information from R & D. Examples include definition of new material properties of graphite epoxy composites, or new computer designs that provide the necessary computational speed. This is the quantification of the creative solutions, the modeling, the numerical analysis portion of the design process.

Solution Evaluation

When the various proposed solutions have numbers associated with them, they have been quantified to some degree. A judgment must be made as to which solution is the optimum one. Economic analysis is included; manufacturing costs and marketing strategies are all considered. We can determine the power requirements of various compressor designs, for instance, but evaluating design involves more than selecting the most efficient compressor. Our evaluation must include how the compressor fits into the company's product line, whether it competes with others or achieves aesthetic value, and other items. From such an evaluation list a solution is selected.

Problem Resolution/Solution Implementation

At this point in the design process the problem statement must be reviewed to assure that what has been created solves the design problem. Additionally, the solution must be manufactured. Do the technologies exist to implement the solution? If the machining tolerances of the present tools cannot meet the requirements of the solution and new tools cannot be purchased, a new solution must be considered. The problem is not resolved. Assuming that such difficulties

The translation of a design on paper to the actual device can be surprising. This coal pulverizer is several stories high, and effects that can be ignored in small systems become important in large ones like this. (Courtesy of Public Service Electric and Gas Company)

are not present, then the manufacturing process must be implemented and the product produced or the project initiated. For large-scale projects, such as creating a new antibiotic, the initial solution will be a laboratory model which is scaled up to a pilot plant. New problems need to be solved at this level, finally resulting in a full-size production facility. This type of undertaking requires the coordination of many people, from trades workers to research scientists, for its successful implementation.

8.9 ADDITIONAL CONSIDERATIONS IN THE DESIGN PROCESS

Several other situations affect the design of a product or process. We bring to the problem we are trying to solve our own value system. These value systems shape how we see the world and interpret what is important on a subjective level.

Engineers must be aware of not only their cultural, regional, and social beliefs, but also those of the customer. This sensitivity is one of the reasons for taking courses in social sciences and humanities, to increase your breadth and depth of understanding of other people.

In line with this a variety of human factors should be considered in the design of a product. These ergonomic considerations deal with the human/machine interaction. Such factors as the relation of the equipment to a person's size, the ease of moving a dial or adjusting a CRT on a computer, the selection of colors that are harmonious or jarring, depending on the device's function, fit within this category. Often the arrangement of equipment, so it minimizes crowding of people and assures a safe working environment, physiologically and psychologically, is a concern of design teams.

Most often engineers work as part of a team. Usually the product that is being designed is too large for one person to design completely, so a team must work together, each having individual and collective responsibilities. Consider the design of a home refrigerator. Various components and aspects must be considered once the size is determined. These include the compressor specification, including sealing of the unit, piston size, lubrication and bearings, and refrigerant selection; the structure of the motor that drives the compressor, including the stator, rotor, poles, lubrication, and bearings; the size of piping to the condenser and to the evaporator; the size of the evaporator and condenser; the control system that actuates the motor/compressor; autodefrost capability; light and aesthetic considerations. This list could be expanded further. The point is that various individuals work on each of the components or subsystems, and then share information as they design the entire refrigerator system. Very often one engineer depends on the output from another subsystem, designed by another engineer, before he or she can complete the design of a component.

In other types of design projects, such as in large devices like ships, the components are already designed and available from manufacturers. The manufacturers provide catalogs that contain the specifications and operating characteristics of heat exchangers, motors, engines, radar, and generators. The design team in this case takes these known components and creates new systems. The engineer must be sure that the characteristics of the subsystems are compatible with one another. For instance, the operating speed of a motor should not be at the critical speed of the driven pump's rotor, or it would vibrate very severely. It should be apparent that communication among design team members is very important to design and development of successful products.

The design process should not end when the product is produced. There will be redesign, improvements made in light of consumer comments, new technologies that can be incorporated. The United States has taken the lead in creating inventions, winning Nobel prizes. It has lost when it comes to innovation, those improvements to initial products that make them less costly, better quality, a better value. Table 8.8 illustrates a list of devices that were invented in the United States and the market share they lost. To regain the lead in technological prowess, not only must engineers in the United States invent new products, but also the companies and the engineers must keep improving

TABLE 8.8 INVENTION/INNOVATION
The Market Share History of Several Products

U.S.-invented technology	1987 market in millions	U.S. producers (%)			
		1970	1975	1980	1987
Phonographs	$ 630	90	40	30	1
Color TV	14,050	90	80	60	10
Audiotape recorder	500	40	10	10	0
Videotape recorder	2,895	10	10	1	1
Machine tool centers	485	99	97	79	35
Telephones	2,000	99	95	88	25
Semiconductors	19,100	89	71	65	64
Computers	53,500	NA	97	96	74

Source: U.S. Commerce Department.

these products. Innovation is mandatory in maintaining a competitive edge with foreign manufacturers.

Developing high-quality products requires more than instituting quality-control checkpoints. It requires a corporate attitude summarized in the term "total quality assurance" (TQA), a concern about developing and maintaining high-quality products and services throughout the organization. Recent studies have indicated that quality is a high priority for today's consumers, corporate and individual. Ten years ago in one survey, only 30 to 40 percent of the respondents considered quality as important as price; that figure has risen to 80 percent today. Companies have also found that quality improvement and cost cutting can go hand in hand. Having products perform according to specification reduces scrap (out-of-spec parts), repair work on returns, and warranty work in the field. A large manufacturer reported these costs to be 20 percent of revenues before instituting an extensive TQA program.

There are two basic approaches to quality improvement, benchmarking and employee involvement. Benchmarking means searching for the best product or service in the industry and using that as a basis for judging your own company's performance. The concept of TQA extends beyond manufacturing a product, into distribution operations and billing as well. Once goals for improving service and products have been objectively determined, employees need to be involved in achieving the objectives.

Employee involvement can be expensive in the short term. It requires training in fundamental subjects like quality improvement, where workers learn what happens to various components when they finish with them and whether their fellow employees who use the components are satisfied with them. In addition workers conduct meetings and jointly solve problems at a level where the problems occur. Having workers solve problems at a lower level than is typical in many corporations can require management reorganization, usually to the detriment of some middle managers. Changing corporate attitudes does not

Increased use of computer-controlled machinery has improved the quality of manufactured components. These engineers are discussing the output from the computer numerically controlled machines in the foreground. (Courtesy of Pall Corporation)

come easily, a reason that giving more than lip service to TQA is difficult at times.

Companies who are willing to undertake a TQA program follow four steps. First, a variety of charts are created, plotting the performance of various sectors of the manufacturing process. The results from one of these charts tell when a machine should be adjusted, before its parts fail to meet specifications, rather than waiting for the out-of-specification condition to arrive. As a machine ages, its performance will deteriorate. By continually monitoring the output, you can have the machine serviced before the output deteriorates to the point of not meeting specifications.

Second is the design, or redesign, of products that are simpler to assemble. In general the reliability of a simply designed product will be greater than a more complex design. Third, the number of suppliers is reduced so as to encourage a closer relationship with them. By being a larger and more frequent customer your quality standards have a greater chance of being accepted. For instance, when Xerox adopted a TQA program in the manufacture of its copiers, it reduced suppliers from 5000 to 400. To maintain a good working relationship with Xerox, suppliers must maintain their expected quality standards.

Last, statistical methods, based on the works of Genichi Taguchi and used in the control of manufacturing operations, are incorporated to minimize disruptions in these operations. For instance, if a process is temperature-dependent and the temperature in the plant cannot be adequately controlled, the Taguchi method would indicate that a temperature-independent process be developed. Don't continue with the existing one. This methodology also allows for a comparison of other materials and methods of manufacture than those currently employed. Checking for improvement may be made on a continual basis.

REFERENCES

1. *Machine Design*. A weekly journal that contains a multitude of electrical, electronic, and mechanical devices.
2. Yeshurum, Sharaga. *The Cognitive Method*. National Council of Teachers of Mathematics, Reston, VA, 1979.
3. Van Amerangen, C. *The Way Things Work*. Simon and Schuster, 1971.

PROBLEMS

8.1 As president of an engineering society at your school you have decided to celebrate winning the regional design contest. Fifty people are being invited; those you know will receive an invitation postcard, costing $0.18 in postage each, and those you do not know will receive a letter, costing $0.25 in postage. If the total postage paid is $11.10, determine the number of letters and postcards sent.

8.2 Determine the significant figures in each of the following numbers.
   a) 59.21   **b)** 642.01   **c)** 1.7002   **d)** 3.8
   e) 4.31×10^2 **f)** 8.3201×10^{-2} **g)** 0.009 00   **h)** 9100

8.3 Write the following numbers in scientific notation.
   a) 561.92   **b)** 8.31   **c)** 0.0396   **d)** 3101.25
   e) 729 034   **f)** 93 000   **g)** 0.0004   **h)** 43.10

8.4 Convert the following numbers from scientific notation to decimal notation.
   a) 8.342×10^2 **b)** $1.693\ 21 \times 10^{-3}$ **c)** 9.8316×10^{-1}
   d) 9.3×10^4 **e)** 7.029×10^3 **f)** 3.0×10^{-2}

8.5 Perform the following additions, giving the answer to the correct number of significant figures.

a) 3.892	**b)** 0.0695	**c)** 4.50×10^3
0.2305	3.4509	1.450×10^3
25.9	20.021	2.31×10^2

d) 0.854	**e)** 7.3×10^{-4}
0.0094	3.65×10^{-1}
0.021	5.9×10^{-3}

8.6 Perform the following multiplications and divisions, giving the answer to the correct number of significant figures.
   a) $(0.289)(0.067) =$ **b)** $(15.79)(0.5731) =$
   c) $(90.620)(-12.21) =$ **d)** $(8957.1)(0.541) =$
   e) $(6.893 \times 10^{-2})/(0.035) =$ **f)** $(7.346 \times 10^{-3})/(6.456 \times 10^2) =$

8.7 Convert the following.
  a) 75°F to C   **b)** −40°C to F   **c)** 132°F to K   **d)** 10°R to K
  e) 10°R to C   **f)** 50°C to F   **g)** 50°C to R

8.8 Convert the following.
  a) 50 million gal/day to m^3/s **b)** 500 m/min to miles/h
  c) 2000 lbm to kg   **d)** 20 gal/min to l/s
  e) 88 ft/sec to km/h

8.9 An electric utility charges its customers 12 cents per kilowatt hour (kwh) of usage up to 2000 kwh and 15 cents per kwh above 2000 kwh. A business uses 25 kW per day for 20 days per month and 5 kW for the remaining 10 days. What is the monthly electric bill?

8.10 An electric power generation station uses ten tons of coal per hour. If the ash content of the coal is 12%, how many pounds of ash are produced each day?

8.11 In the equation $\dot{m} = \rho A v$, $\dot{m}$ is the mass flowrate, ρ is the density, A is the area, and v is the velocity. Find $\dot{m}$ in lbm/min if ρ is 60.0 lbm/ft^3, A is 288 in^2, and v is 2 ft/sec.

8.12 In the equation $F = ma$, F is 100 N and m is 5 kg. Find the acceleration in m/s^2 and ft/sec^2.

8.13 The time required for a pendulum to complete one complete swing is $t = 2\pi(L/g)^{0.5}$, where L is pendulum length and g is gravitational acceleration. If the time is measured at two seconds and the length is one meter, determine the value of g in m/s^2. Convert the length to feet and compute the gravitational acceleration in English units (you may need g_c in the equation; perform a unit balance to check).

8.14 You have been appointed chair of a student committee that must solicit opinions from students, faculty, administrators, alumni, and corporate executives regarding a major engineering project. The time frame for the committee's final report is six months. During this time you must organize the committee, hold committee meetings, solicit opinion fro m the various constituencies, consolidate information, and write the final report. Determine the activities necessary to complete this project, and construct a Gantt chart for them.

8.15 Using the multicriteria methodology, develop a model for evaluating the following.
  a) the purchase of a new automobile
  b) the decision to live on or off campus
  c) changing universities
  d) changing from an engineering major to a nonengineering one

8.16 Repeat the decision analysis for the oil company in the situation where the competitor offers $400,000, the cost of drilling is $1,400,000 and the probabilities are 25% for a dry hole, 35% for gas only, 25% for gas and oil, and 15% for oil. Determine the decision tree for the analysis and what your recommendation would be and why.

8.17 List questions that must be resolved in defining the following problem statements.

   a) Develop a new ski binding.

   b) Develop an automatic bicycle gearshift.

   c) Develop an improved wheelchair.

   d) Develop a baby stroller/carriage.

   e) Develop an automatic cash machine for banks.

8.18 Create solutions for the design problems in Problem 8.17, subject to the constraints you have imposed.

8.19 Design a car bicycle rack.

8.20 Design a backpack.

8.21 Investigate the limitations of a battery-powered electric car, and create a preliminary design of one.

8.22 You have just finished working in an office during the summer, and as part of your job you had to seal 100 envelopes a day. Tired of a sticky tongue, you decide to create an automatic envelope sealer. Describe the preliminary design.

8.23 As a bicyclist you often have small packages to carry that are too large for the typical carrying racks. You also do not like to wear a backpack. Design a bicycle rack that will carry packages up to one cubic foot.

8.24 As a concerned citizen and an advocate of recycling of waste you realize one of the problems with recycling programs is the lack of a mechanism to sort refuse components. Assume the following components must be separated: ferrous materials (tin cans), glass, plastic, and aluminum. Create a separation system.

ELECTRICAL SYSTEMS

■ To learn about the elements of a circuit.

■ To apply Kirchhoff's laws to direct-current circuits.

■ To analyze transient direct-current circuits.

■ To design waveforms.

■ To understand how alternating-current circuits work.

(Photo courtesy of Thomas Ives.)

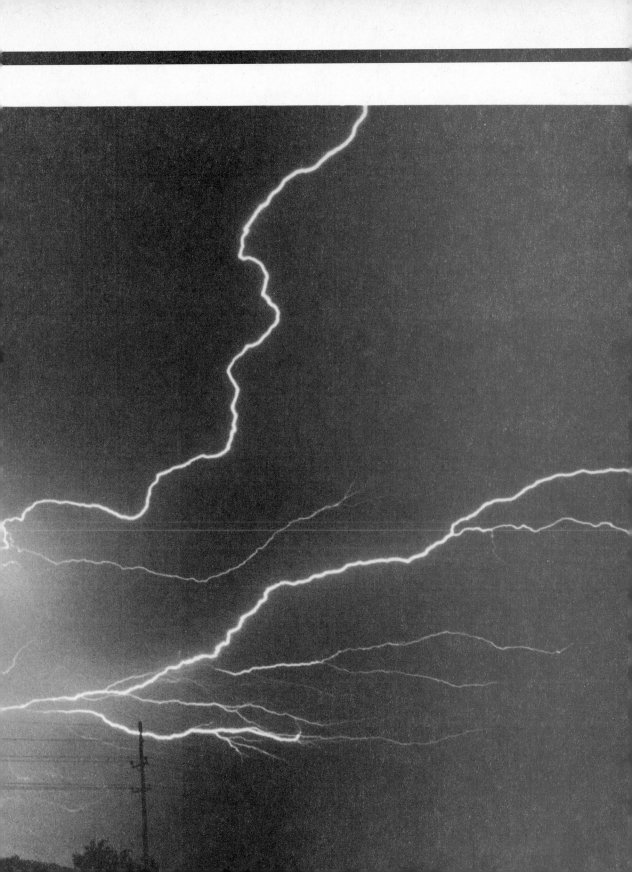

Engineers have created the systems that run the modern world, powering transportation networks, lighting homes, allowing worldwide communication, and speeding the healing of injuries. Electrical engineers are a major force in the evolution of our industrial society.

9.1 INTRODUCTION

In today's world, we depend heavily on electrical systems for support of our lifestyles and livelihoods. Think for a moment about all the features of our daily life that involve electrical or electronic components: waking in the morning to an alarm clock, listening to the radio or watching television, starting an automobile, turning on a light to use a calculator. In every case an electrical system is involved, created, and developed by electrical engineers.

Electrical engineering is a broad term encompassing electrical power generation, distribution, and use and electronics, the study of electron-charge motion through semiconductors and in a vacuum. Electronics engineering, which is limited to applications such as computers, integrated circuits, communications, fields, and waves, is considered a subset of electrical engineering.

All electrical circuits can be mathematically modeled using one or more circuit elements, resistance, inductance, capacitance, voltage, and current sources. Figure 9.1 shows the symbols used to designate the elements. The flow of electrical charge which requires work to move it from one point to another is called current, a fundamental unit measured in amperes (A). One ampere is equal

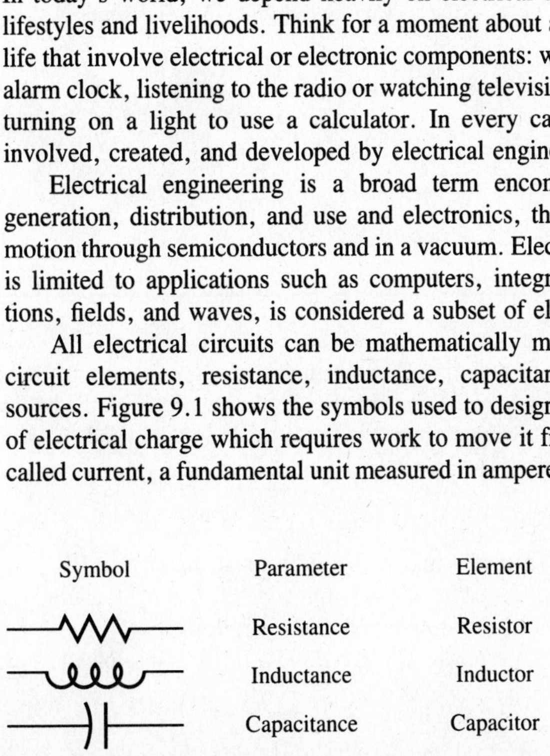

Symbol	Parameter	Element
—\/\/\—	Resistance	Resistor
—ꞶꞶꞶ—	Inductance	Inductor
—)|—	Capacitance	Capacitor

FIGURE 9.1 Electrical circuit elements.

A printed circuit board after wave soldering. In wave soldering the circuit board is positioned close to the liquid solder surface in a tank. A small wave of solder is generated and passes across the circuit board, soldering the circuit connections. There is no trapped air or voids as would occur if the circuit board were placed on the liquid surface. (Courtesy of Grumman Corporation)

to the flow of one coulomb per second. A volt (V) is the change in electrical potential between two points if one joule of work is done in moving one coulomb of charge from one point to another.

9.2 RESISTANCE

The definition for resistance is derived from Ohm's law, named after Georg Simon Ohm, a German physicist who discovered that the voltage change (V) across a resistor (measured in volts) is equal to the product of the electric current (i) flowing through the resistor in amperes and the resistance (R) measured in ohms (Ω):

$$V = iR. \qquad\qquad 9.1$$

Equation 9.1 may be solved for resistance, $R = V/i$. Since this is a law, it cannot be proven, only corroborated by experimental evidence. As with most empirical laws, there are some minor exceptions, but none of note at this level.

The power (P) in watts (W) that is dissipated in the resistor is

$$P = Vi \qquad\qquad 9.2$$

which, if combined with Equation 9.1, yields

$$P = i^2R. \qquad\qquad 9.3$$

In any resistor, the power loss is referred to as the i-squared R loss. All conductors have some electrical resistance, so there are power losses associated with current flow through virtually all conductors. In electrical distribution systems this is very important; if you wish 100 MW of power to be transmitted 100 miles, then you may need to provide 110 MW at the beginning of the transmission line to receive 100 MW at the end.

9.3 STEADY-STATE DC CIRCUITS

In steady-state DC circuits the voltage and current values do not change with time. This is typified by a battery and resistor circuit, shown in Figure 9.2. In the battery a chemical reaction converts chemical energy into electrical energy. This process moves electrons, negatively charged, from material at the positive terminal (anode) to the negative terminal (cathode), creating an electric potential. If a load, in this case the light bulb acting as a resistor, is connected between the terminals, the electrons will flow from the negative terminal to the positive terminal. The diagram indicates that the conventional current flow, which is hereafter called the current flow, is in the opposite direction. This is the direction in which positive-charge carriers would move if the current moved from the positive to negative terminal. The movement of positive carriers is important in semiconductor devices, where these carriers are called holes. The reason for this inverted direction for current flow is historical. Benjamin Franklin proposed that electrical current was the flow of positively charged particles. Though later proven wrong, the proposal remains and is accepted practice.

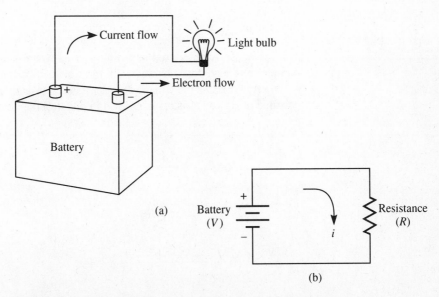

FIGURE 9.2 (a) A battery/resistor circuit and (b) its circuit diagram.

Kirchhoff's Laws

Two laws discovered by Gustav Kirchhoff in the mid-1800s are extremely valuable in analyzing electrical circuits. The first is frequently called Kirchhoff's voltage law:

> The sum of the voltage rises around a closed loop in a circuit must equal the sum of the voltage drops.

The second of Kirchhoff's laws is sometimes called Kirchhoff's current law:

> The sum of all currents into a junction (node) must equal the sum of all currents flowing away from the junction.

The first law assists us in the analysis of resistors connected in series, as shown in Figure 9.3. Three resistors are connected to a DC power source. The circuit has one voltage rise, the battery with voltage V, and three voltage drops, the iR drops across each resistor. Expressing this in an equation yields

$$V = iR_1 + iR_2 + iR_3. \qquad 9.4$$

The current flow i is the same for each resistor, hence

$$V = i(R_1 + R_2 + R_3) = iR_{eq} \qquad 9.5$$

where R_{eq} is the single equivalent resistance that could replace all three individual resistances. An equivalent resistance is equal to the sum of the individual resistances.

$$R_{eq} = \sum_i R_i \qquad 9.6$$

The symbol $\sum_i$ denotes a summation of the subscripted variable R_i. The equivalent resistance is often used in testing circuits where you are interested in the total resistive load characteristics, not the resistive load of individual resistors.

In series circuits the voltage drop across each resistor will vary according to the value of R. For resistors connected in parallel, the voltage drop across each

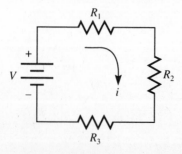

FIGURE 9.3 Resistors connected in series.

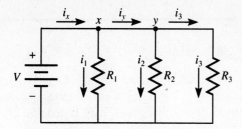

FIGURE 9.4 Resistors connected in parallel.

resistor is the same and equal to the battery voltage (Figure 9.4). From Kirchhoff's voltage law we determine

$$V = i_1 R_1 = i_2 R_2 = i_3 R_3. \qquad 9.7$$

From the current law for junctions x and y we find

$$i_x = i_1 + i_y \qquad 9.8a$$

and

$$i_y = i_2 + i_3. \qquad 9.8b$$

Combining Equations 9.8a and 9.8b,

$$i_x = i_1 + i_2 + i_3. \qquad 9.9$$

This makes sense physically in that the current leaving the battery equals the sum of the individual resistor currents. The individual currents may be found by applying Ohm's law to each resistor:

$$i_1 = \frac{V}{R_1}, \qquad i_2 = \frac{V}{R_2}, \qquad i_3 = \frac{V}{R_3}. \qquad 9.10$$

Substituting Equation 9.10 into Equation 9.9,

$$i_x = V\left(\frac{1}{R_1} + \frac{1}{R_2} + \frac{1}{R_3}\right). \qquad 9.11$$

Notice that the term in parentheses is the inverse of the equivalent resistance R_{eq} for a parallel circuit and that Equation 9.11 is an expression of Ohm's law:

$$\frac{1}{R_{eq}} = \frac{1}{R_1} + \frac{1}{R_2} + \frac{1}{R_3}. \qquad 9.12$$

This may be reduced to a single term:

$$R_{eq} = \frac{R_1 R_2 R_3}{R_1 R_2 + R_2 R_3 + R_1 R_3}. \qquad 9.13$$

We can use the previously developed equations in simplifying circuit diagrams.

E X A M P L E

9.1

Find the equivalent resistance of the circuit shown in Figure 9.5a.

SOLUTION:

Start with the resistance(s) furthest from the voltage source. In Figure 9.5a the 15 Ω and 20 Ω resistors are in series and combine to form one 35 Ω resistor, as shown in Figure 9.5b. At this point the 10 Ω and 35 Ω resistors are in parallel and can be combined to give an equivalent resistance.

$$\frac{1}{R_{eq}} = \frac{1}{10} + \frac{1}{35}$$

$$R_{eq} = 7.78 \ \Omega$$

The circuit diagram now is shown in Figure 9.5c. The two remaining resistors are in series and can be added, as shown in Figure 9.5d, yielding an equivalent resistance for the circuit of 12.78 Ω.

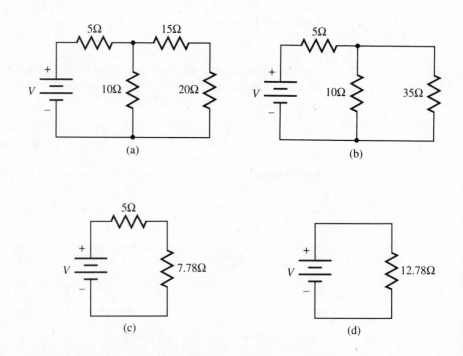

FIGURE 9.5 Circuit diagrams for Example 9.1.

9.4 APPLICATIONS OF RESISTIVE CIRCUITS

Two applications that we will examine for purely resistive circuits are the variable voltage divider, also known as a potentiometer, and the Wheatstone bridge. A potentiometer is a rheostat, a device whose resistance can be varied (Figure 9.6). One of the leads of the resistive material is attached to ground, with leads of the wiper and the other end connected to the voltage source. Figure 9.7 illustrates a potentiometer in a circuit; the potentiometer's wiper has been connected to a loudspeaker. The variable resistance changes the voltage supply (v) across the speaker, thereby changing the volume. The lower case v is used for the variable voltage supply of the music coming from the amplifier. In this situation the voltage is applied across the potentiometer's total resistance, with the loudspeaker connected between the wiper and ground. In Figure 9.7a, the speaker receives only 20 percent of the voltage drop, and the volume is low; in Figure 9.7b the wiper—the volume knob on the stereo—has been turned to allow 80 percent of the voltage drop to occur across the speaker, and the volume

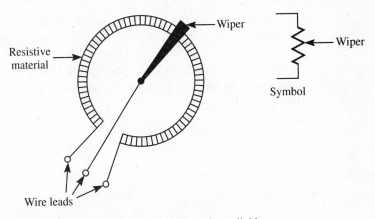

FIGURE 9.6 A potentiometer, or variable voltage divider.

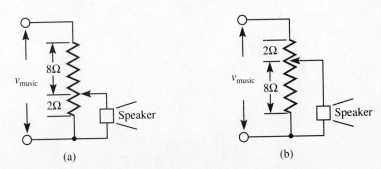

FIGURE 9.7 The potentiometer used as a variable voltage divider.

is higher. Variable resistance potentiometers are used as volume and tone controls in radios and stereophonic equipment and as contrast and brightness controls on televisions.

The Wheatstone bridge is a special case of variable resistance, included in a parallel series of resistors, that can measure very small values of resistance such as those found in strain gages. Strain gages are used in testing the strength of materials to find the load a material is subjected to and the effect of that load on creating a strain in the material. Figure 9.8 shows a Wheatstone bridge with current flows in the resistors for all the branches. There are resistances in each of the arms and a meter with a resistance R_m. In practice R for one arm is the unknown, and the bridge is balanced to find the unknown resistance, frequently the strain. Let's consider that the bridge is balanced; thus i_m, the current through the meter, is zero. The current $i_1 = i_2$ and $i_3 = i_4$. Furthermore, the potential across the meter must be zero, thus

$$i_1 R_1 = i_3 R_3 \qquad \text{and} \qquad i_2 R_2 = i_4 R_4.$$

Eliminating the i_1 and i_4 from the previous equation and dividing the result yields

$$R_1/R_2 = R_3/R_4. \qquad\qquad 9.14$$

In a Wheatstone bridge R_1 and R_2 may be known resistances, R_3 a variable resistance, R_4 an unknown resistance, and R_m the resistance of a galvanometer, which indicates the current i_m in the unbalanced bridge arm. The variable resistor is adjusted to bring i_m to zero. The final resistance of R_3 is read from a dial attached to it, and the value of the unknown resistance is determined from Equation 9.14.

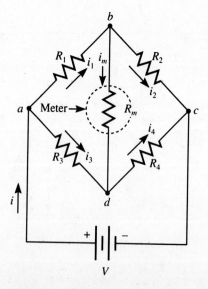

FIGURE 9.8 A Wheatstone bridge circuit.

The capacitors attached to this utility pole change the impedance of the electrical supply, counteracting in part the effects of inductance. (Courtesy of Public Service Electric and Gas)

Another use of a balanced bridge is an alarm system. Consider the circuit in Figure 9.8 with the resistances balanced per Equation 9.14 and in place of the meter an alarm, which activates if current passes through it. Now one of the resistances is a thermistor, a resistor whose resistance changes with temperature. One resistor of the bridge unit is located in a refrigerated space. If the compressor fails and the temperature rises beyond a certain point, allowing a sufficient change in resistance, the current flow will be great enough to trigger the alarm. Note that changes in voltage will not affect the alarm.

The circuits described above are totally resistive and hence are not transient. There is no delay in current or voltage in a resistive circuit; all changes occur instantaneously. However, other circuit elements, capacitance and inductance, have properties that do not delay changes in current and voltage.

9.5 CAPACITANCE

Another circuit element, capacitance, is denoted most frequently as a capacitor which will have a voltage v across it, an electrical charge q stored within it and measured in coulombs, and a property C, its capacitance, measured in farads. The equation defining the relationship is

$$v = \frac{q}{C}$$

9.15

which in essence defines the capacitance. Equation 9.15 is physically interpreted to say that a one-farad capacitor with one coulomb of charge stored in it will have a potential difference of one volt across its terminals. The utility of this relationship becomes apparent when we consider the time variation of charge change. We know that current is measured in coulombs per second; thus

$$i = \frac{\Delta q}{\Delta t}$$

or written as a differential letting the Δ's become very, very small (differentially small),

$$i = \frac{dq}{dt}. \qquad 9.16$$

Assuming that we can apply Equation 9.15 to the time-varying situation, it may be written as

$$\frac{\Delta v}{\Delta t} = \frac{1}{C} \cdot \frac{\Delta q}{\Delta t}$$

on a difference basis. Letting the values become differentially small yields

$$\frac{dv}{dt} = \frac{1}{C} \cdot \frac{dq}{dt}. \qquad 9.17$$

Combining Equations 9.16 and 9.17 yields

$$i = C \cdot \frac{dv}{dt}. \qquad 9.18$$

Equation 9.18 can be used to model the change of voltage and the current flow in a capacitor. Note that when the voltage is constant, the current must be zero.

RC Circuits

Equation 9.18 is not an end in itself and is most frequently encountered in the analysis of circuits, the simplest being the resistance-capacitance circuit, or RC circuit, illustrated in Figure 9.9. In this situation we assume the capacitor is

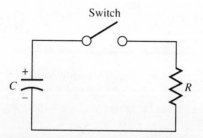

FIGURE 9.9 A simple resistance-capacitance circuit.

charged, and when we close the switch, the current will flow through the resistance R, discharging the voltage. We can determine the form that the discharge process will take, being able to predict what the voltage change with time will be.

At all time intervals the voltage rise across the capacitor must equal the voltage drop across the resistance, thus

$$iR = \frac{q}{C}.$$
9.19

Furthermore, we know that the current can be expressed as charge per unit time, $i = dq/dt$. Substituting this in Equation 9.19 with the caveats that the dq must be negative, the capacitor is discharging, and a positive current flow implies a minus sign ($i = -dq/dt$ in this case),

$$-R \cdot \frac{dq}{dt} = \frac{q}{C}.$$

Combining the same variables on separate sides of the equation,

$$\frac{dq}{q} = -\frac{1}{RC} \cdot dt.$$

This equation must be solved to determine q in terms of RC and t. The solution, which is beyond the scope of this text, is

$$q = Ae^{-t/RC}$$
9.20

where A is an unknown coefficient that must be determined. Let the value of the charge at time zero be q_0. Substitute these values, $q = q_0$ at $t = 0$, into Equation 9.20:

$$q_0 = Ae^0 = A(1).$$

Now we know the value of A and can substitute it in Equation 9.20:

$$q = q_0 e^{-t/RC}.$$
9.21

Voltage, not charge, is what is measured in circuits, so using Equation 9.15 finally provides us with the variation of voltage with time for the discharge of a capacitor:

$$v = q_0/C \cdot e^{-t/RC}.$$
9.22

Figure 9.10 graphs the discharge of a one-millifarad capacitor through a 100-ohm resistor. At time zero, the voltage in the capacitor is q_0/C, and it decreases exponentially with time until its value is zero. Theoretically an infinite amount of time is required, but practically after one-half second the value of the voltage is less than 1 percent of its initial value, essentially zero.

While we are investigating the discharge of the capacitor, another point of interest to electrical engineers occurs when the value of t equals RC. This is referred to as the time constant of the circuit. In this case the exponent in Equation 9.22 is minus one, yielding a voltage value of 0.368 times the initial

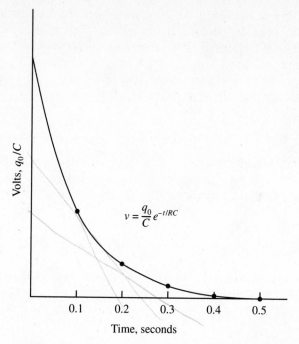

FIGURE 9.10 The voltage change with time of a discharging capacitor.

voltage; the capacitor has lost approximately 63 percent of the initial charge. In Figure 9.10 the time constant of the capacitor is 0.1 second; hence the capacitor lost 63 percent of its voltage in one-tenth of a second, or 100 milliseconds.

Waveforms

To illustrate the concept of a simple waveform, consider a car's turn signal, diagrammed in Figure 9.11a. A switch is opened and closed by a timer. For a purely resistive circuit, the current reaches its steady-state value without delay, and the current waveform for a one-second timer is shown in Figure 9.11b. Thus, the turn-signal light will be on for a second, off for a second, on for a second, and so forth, and the current waveform is a square wave.

The value of the time constant lies in its use in helping us model various waveforms. More complicated waveforms are often necessary to create in electronic circuits, with the sawtooth waveform being one of the more common types. Figure 9.12 shows a sawtooth waveform. A problem in electrical engineering design is to create a circuit that will model this waveform.

Figure 9.13 shows an RC circuit that can be used in the solution. When the switch is at point *a*, the circuit charges the capacitor from the battery. This causes a buildup of voltage in the capacitor, which follows an exponential function, to be determined. When the switch is moved to *b*, the capacitor is shorted,

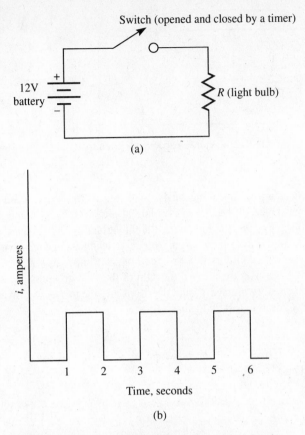

(a)

(b)

FIGURE 9.11 (a) The circuit diagram for a turn signal. (b) A current square waveform for the circuit.

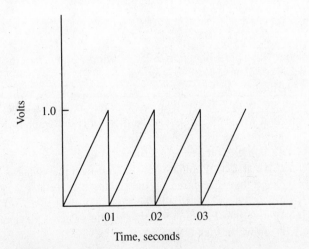

FIGURE 9.12 A sawtooth waveform.

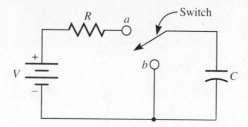

FIGURE 9.13 The resistance-capacitance circuit that creates the sawtooth waveform.

discharging instantaneously. We reach this conclusion by noting that the resistance in the circuit is minimal and may be assumed to be zero. Thus, the discharge process models a straight vertical line. Let's now determine the voltage change during the charging process, when the switch is at point a.

From Kirchhoff's voltage law, we know that at any time the sum of the voltage rises must equal the sum of the voltage drops. The voltage rise is due to the battery voltage V, and the drops are due to the iR drop across the resistor and q/C drop across the capacitor:

$$V = iR + q/C. \qquad 9.23$$

Furthermore, $i = dq/dt$, where dq is positive as the capacitor is being charged. Thus,

$$V = R \cdot \frac{dq}{dt} + q/C \qquad 9.24$$

which may be integrated to yield

$$q = CV(1 - e^{-t/RC})$$

or

$$v = V(1 - e^{-t/RC}). \qquad 9.25$$

Figure 9.14 plots Equation 9.25. How do we get from this curve to a sawtooth wave? The curve rises very rapidly at $RC < 1$; we could say that the curve is more linear in its beginning portion. If we allow the voltage to rise briefly, then open the switch, a sawtooth waveform is created. However, we must choose values of RC and V such that the model fits the graph in Figure 9.12.

The process uses trial and error. The value of V is assumed, the lower the better, as high voltages are detrimental to electronic circuits and require more expensive grades of components to withstand the voltage. The maximum value of voltage that is required by the sawtooth wave is 1.0 volt, which we might assume occurs at one-half the time constant; thus RC is 0.02 seconds. Substituting into Equation 9.25 yields the value of the applied voltage V in the circuit shown in Figure 9.13:

$$1.0 = V(1 - e^{-0.01/0.02}); \qquad V = 2.5 \text{ volts.}$$

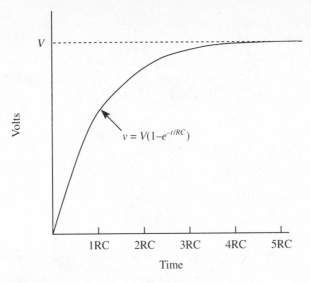

FIGURE 9.14 The graph of Equation 9.25.

If the battery voltage is 2.5 volts and RC is 0.02, how can we determine if the circuit will function as shown in Figure 9.12? Check the midpoint. At $t = 0.005$, the voltage should be 0.5 volts if the waveform is linear. Applying Equation 9.25 again yields

$$v = 2.5(1 - e^{-0.005/0.02}) = 0.55 \text{ volts.}$$

This is not a terrific match, so a few iterations are helpful. In this process we select RC, and knowing $v = 1$ volt at 0.01 seconds, we determine the battery

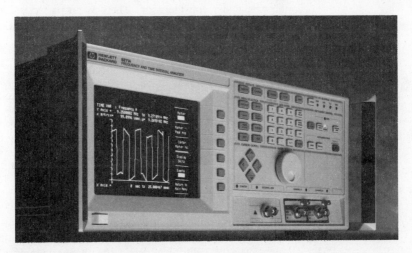

This frequency and time interval analyzer provides a three-dimensional analysis of complex signals. This equipment is used to test microwave and radar signals. (Courtesy of Hewlett-Packard)

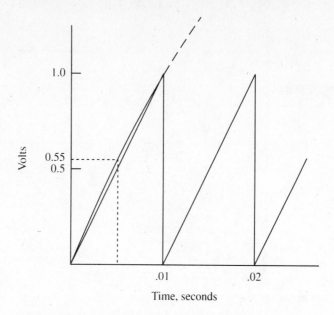

FIGURE 9.15 The graph of Equation 9.25 and the sawtooth waveform that it will, in part, model.

voltage for this selection. The deviation at the midpoint is a measure of the linearity of the sawtooth wave. Continuing, if RC is 0.04, V is 4.5 volts and the midpoint voltage is 0.53 volts; if RC is 0.06, V is 6.5 volts and the midpoint voltage is 0.52 volts. Figure 9.15 illustrates the sawtooth waveform that we have created. You as the design engineer must decide what accuracy is required for the application under consideration. There is no perfect solution in this situation, only more-accurate solutions, which are more expensive.

9.6 INDUCTANCE

Another circuit element is inductance, L, which delays the change of current and is related to voltage and current change by the following equation:

$$v = L \cdot \frac{di}{dt},$$ 9.26

where L is measured in henries (H) and is defined by Equation 9.26.

A simple RL circuit is illustrated in Figure 9.16. We know that the resistance delays neither the current nor the voltage change in the circuit when the switch is closed, but the inductance delays the change in current. Let's determine the mathematical relationship for this.

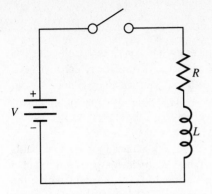

FIGURE 9.16 A simple resistance-inductance circuit.

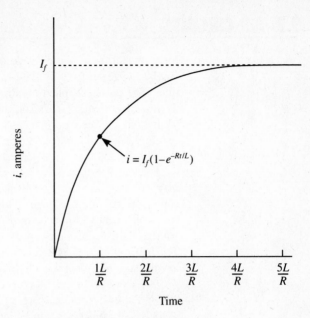

FIGURE 9.17 The graph of Equation 9.29.

The switch is closed, the current starts to flow, and the voltage rise from the battery must equal the voltage drops across each circuit element (this is analogous to the RC circuit derivation):

$$V = iR + L \cdot \frac{di}{dt}.$$ 9.27

The voltage drop across the resistor is iR and across the inductor is $L \cdot di/dt$, per Equation 9.26. Separating the variables and combining them on each side of the equation yields

$$\frac{di}{\frac{V}{R} - i} = \frac{R}{L}dt.$$ 9.28

The term V/R represents the final steady-state value of the current I_f flowing through the circuit as determined from Ohm's law. This equation is of the same form as that of the RC circuit, and the solution of it yields

$$i = I_f(1 - e^{-Rt/L}).$$ 9.29

Inspection of Equation 9.29 shows that L/R is the time constant of the circuit, the time it takes for the current to reach 63 percent of its final value, I_f. Figure 9.17 plots Equation 9.29. Thus, RL circuits may be used to create various current waveforms, as the RC circuit was used for voltage waveforms.

9.7 AC CIRCUITS

Most of the discussion of electrical engineering has concerned itself with DC circuits, where the steady-state values of current and voltage are constant with time. The power supplied to our homes, offices, and machines is not DC, but alternating current, or AC. The current and voltage polarities change with time, alternating between positive and negative values. The sine wave, shown in Figure 9.18, is the most common alternating current and voltage waveform.

How is this waveform generated? Figure 9.19 illustrates the principal that is used in electric generators and motors. When a conductor passes through a magnetic field and cuts the lines of magnetic flux, a force develops within the conductor that pushes charged particles to one end of it. If the motion is upward, the particles move to one end; if the motion is downward, to the opposite end. An induced voltage is created across the conductor as a result of this movement. The induced voltage is a function of the magnetic flux density, the conductor length, and the velocity with which the conductor is moved through the field.

AC can also be generated by a loop connected to an external load (Figure 9.20). The two principal parts of a generator are the stator and the rotor. The stator has two functions, supporting the device, the outside casing, and providing the magnetic flux, either through permanent magnets or induced magnets. Figure 9.20 shows only the magnetic function of the stator. The rotor shown is a single curved piece of wire, whereas actually it consists of many strands of wire supported on a metal shaft (rotor). The rotor revolves about its axis, and the wire cuts the magnetic field, inducing a voltage. Because the conductor is connected to a load, current flows. Mechanically, the wires terminate at one end of the rotor, and slip rings join the wire ends. Pressing against the slip rings, external to the rotor, are carbon brushes through which the current flows to the load, for

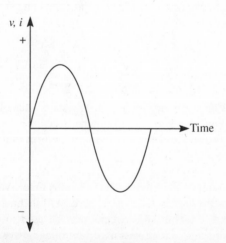

FIGURE 9.18 A sine wave.

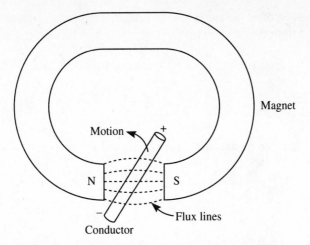

FIGURE 9.19 Voltage is induced in a conductor when it moves through a magnetic field.

example, your stereo. Other generators may have a stationary armature and rotating magnets, but the principle for creating induced voltage remains the same.

Because the wire (armature) cuts through the field in one direction, then the other, and rotates at a constantly varying angle to the fixed magnetic field, the induced voltage and current rise and fall in a sinusoidal pattern. The instantaneous voltage and current are represented by the following equations:

$$v = V \sin \theta$$

and

$$i = I \sin \theta.$$

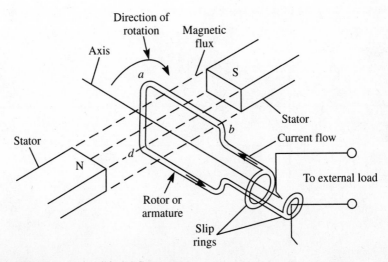

FIGURE 9.20 A simplified AC generator.

Most often the angle θ is not used, but rather the angular frequency ω and time t from the relationship

$$\text{radians} = \frac{\text{radians}}{\text{s}} \cdot \text{s},$$

or

$$\theta = \omega t. \qquad\qquad 9.30$$

Thus the instantaneous current and voltage become

$$v = V \sin \omega t$$

and

$$i = I \sin \omega t.$$

Figure 9.21 illustrates these two equations. In this situation the current and voltage change in the circuit occur simultaneously: they are in phase with each other. This is possible for resistive circuits (or those that act like them). Remember, resistance does not delay the change in voltage or current in a circuit. However, capacitance and inductance do affect the changes of voltage and current, respectively.

The AC waveform for a capacitive circuit is shown in Figure 9.22. The current leads the voltage by 90°, reaching its peak value before the voltage reaches its peak. Should the voltage reach its peak value first, the current would lag behind the voltage. The current is not delayed by the capacitance. This lead of 90°, or $\pi/2$ radians, must be included in its equation, as noted on the graph. An inductive circuit, on the other hand, causes the current to lag behind the voltage by 90°. In Figure 9.23 the current is $\pi/2$ radians behind the voltage in reaching its peak value.

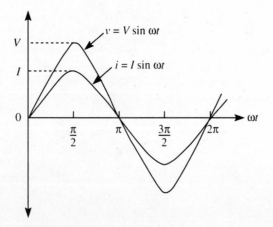

FIGURE 9.21 The AC waveforms for voltage and current in a resistive circuit.

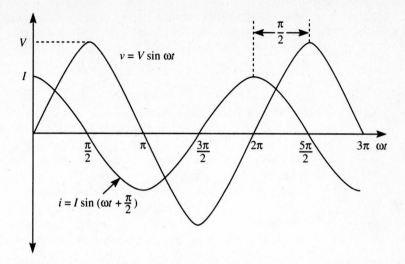

FIGURE 9.22 The AC waveforms for voltage and current in a capacitive circuit.

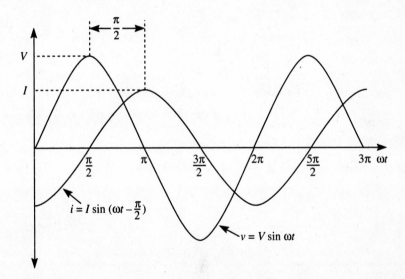

FIGURE 9.23 The AC waveforms for voltage and current in an inductive circuit.

The sinusoidal functions for current and voltage continue in a repetitive fashion for increasing values of ωt. Rather than using ωt as the variable, it is more common to use frequency, the number of complete cycles that occur per second. The frequency, f, and time of one cycle, T, are related by

$$T = 1/f,$$

9.31

where $1/f$ represents the time for one cycle to occur. Since

$$T = 2\pi/\omega,$$

therefore,

$$\omega = 2\pi f. \qquad\qquad 9.32$$

The frequency for household use in the United States is 60 cycles per second, or 60 hertz. In Europe domestic electricity is 50 hertz (and 220 volts, rather than the 110 volts encountered in the United States). All circuits contain inductance, capacitance, and resistance, causing the current to lag or lead the voltage by some degree, usually less than 90°. By adding more inductance or capacitance to the circuit, the phase difference between the current and voltage can be reduced. While the importance of this will not be explored here, a circuit loses power because of the phase difference. Many industrial facilities will add inductors or capacitors, usually the latter, to their electrical networks to reduce the phase difference, their power consumption, and hence the utility bill.

REFERENCES

1. Fisk, J. *Fundamentals of Circuit Analysis*. Macmillan, New York, 1987.
2. Irwin, J. D. *Basic Engineering Circuit Analysis*. 3rd ed. Macmillan, New York, 1989.
3. Kemper, J. D. *Introduction to the Engineering Profession*. Holt, Rinehart and Winston, New York, 1985.

PROBLEMS

9.1 Find the equivalent resistance of the following circuit.

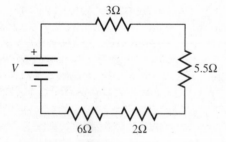

9.2 Find the equivalent resistance of the following circuit.

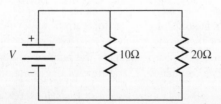

9.3 Find the equivalent resistance of the following circuit.

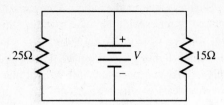

9.4 Find the equivalent resistance of the following circuit and the current through each resistor.

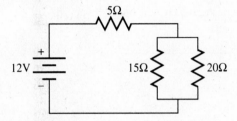

9.5 The filament in a flashlight bulb has a resistance of 40 Ω. The battery voltage is 6 V. Determine the current flow.

9.6 The maximum current flow from a 1.5-V battery is 45 mA. What is the minimum size resistor that can be connected to it?

9.7 Four 5-Ω resistors are wired in parallel.
a) What is the total resistance of the circuit?
b) If they are wired in series, what is the total circuit resistance?

9.8 The equivalent resistance of a three-resistor series circuit is 30 Ω. If the three resistors, each of the same value, are connected in parallel, what is the equivalent circuit resistance?

9.9 Find the equivalent resistance of the following circuit.

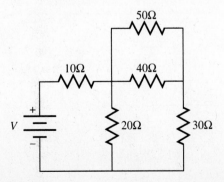

9.10 Your car radio is broken, so you are using a 9-V transistor radio that uses 30 mA of current. Being an engineering student, you wish to conserve the radio's battery and run the radio off the car's 12-V battery. A resistor must be placed in series with the radio to reduce the car's voltage to that of the radio; what is its value? What power does the transistor radio dissipate?

9.11 Many home lighting circuits have 15-A circuit breakers with a voltage supply of 110 V. How many 100-W light bulbs may be placed in parallel in the circuit before the breaker trips?

9.12 You are using a 1250-W hairdryer on a 15-A, 110-V circuit. Your younger sister comes in and turns on a 350-W stereo and a 100-W light, also in parallel on the same circuit. Does the circuit breaker trip?

9.13 What is meant by "balance" when referring to a bridge? How is it accomplished?

9.14 Are the following bridges balanced?

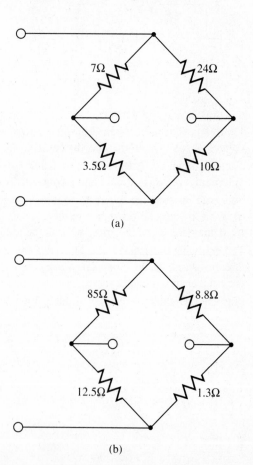

(a)

(b)

9.15 A Wheatstone bridge, as illustrated in Figure 9.8, is used to determine the value of R_1. The bridge is balanced, and R_3 reads 137.5 Ω. R_2 and R_4 are interchanged, the bridge is balanced, and R_3 now reads 167.9 Ω. Determine R_1.

9.16 In a Wheatstone bridge, the resistances R_3 and R_4 are equal. An unknown resistance R_1 is connected to the bridge and the bridge balanced, yielding a value for R_2 of 400 Ω. What is R_1's resistance?

9.17 A capacitor of unknown capacitance has the following parameters when located in a circuit: current flow of 0.2 mA, instantaneous voltage change of 200 V/s. Determine the capacitance in farads.

9.18 An inductor of unknown inductance has the following parameters when located in a circuit: voltage of 30 V, instantaneous current change of 3.0 A/s. Determine the inductance in henries.

9.19 A capacitor of 0.01 F is connected to a 110-V supply. What charge is stored in the capacitor?

9.20 For discharging a capacitor in an RC circuit, assume that the resistance is 10,000 Ω, the capacitance is 12 mF, and the initial voltage is 12 V. Sketch the decay curve. Determine the time constant of the circuit.

9.21 A capacitor in an RC circuit has an initial voltage of 150 V and a capacitance of 12 mF; the resistor is 1 megaohm. Determine the time required to reduce the voltage across the capacitor to 5 V.

9.22 For the inductive RL circuit shown in Figure 9.16, determine the time for the current to reach 90 percent of its steady-state value if $R = 2\ \Omega$ and $L = 4H$.

9.23 Design a sawtooth-waveform current circuit where the peak current is 2 A and the time for each sawtooth is 0.1 s. You must determine the value of R/L and voltage necessary to create the waveform and decide when the accuracy is satisfactory. Describe the criteria you apply in this situation. Remember this is an RL circuit.

9.24 Sketch the voltage and current curves when the current leads the voltage by 45°.

9.25 Sketch the voltage and current curves when the current lags the voltage by 45°.

9.26 The voltage in an AC network is given by $v(t) = 45 \sin 2500t$. Determine the maximum value of voltage, the angular velocity (ω), and the frequency.

9.27 The voltage in Problem 9.24 is applied to a 2000-Ω resistor. Determine the equation for the current flowing through the resistor.

10

MECHANICAL SYSTEMS

■ To describe and resolve forces.

■ To determine moments caused by forces.

■ To understand static equilibrium and the free-body diagram.

■ To calculate the strength of materials.

■ To explore materials science.

■ To analyze motion in one dimension.

*(Photo courtesy of The Hulton Deutsch
Collection/Hulton Picture Collection.)*

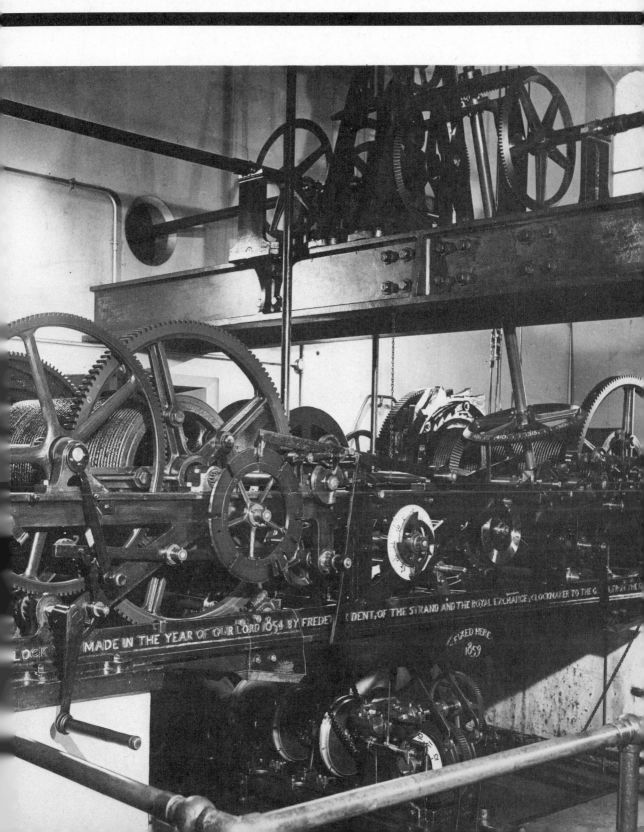

*U*nderstanding the laws of mechanics enables engineers to develop structures that continue to stand even in earthquakes and engines that run on the earth's surface, in the ocean's depths, and in the vastness of space. Engineers design such devices and structures by adding their own creativity to these laws.

10.1 INTRODUCTION

The field of mechanical engineering has two branches, mechanics and energy. In applied mechanics you learn about material properties, forces acting on structures and devices, and the motion of devices, such as rotating machines. This chapter addresses applied mechanics.

One of the first engineering courses you will take is engineering mechanics, the first part of which concerns itself with bodies in static equilibrium. Virtually all majors take this course, as it introduces engineering problem solving and develops your ability to analyze any problem in a logical fashion, using fundamental principles. This skill is transferable to all other engineering courses. The advantages of problem solving in statics, and later in dynamics, is that the situations can be physically visualized. Civil and mechanical engineers are particularly interested in engineering mechanics, as later courses, such as strength of materials and structural analysis and design, depend on it.

10.2 FORCES

The field of mechanics concerns itself with forces acting on bodies at rest—static equilibrium—and forces acting on bodies in motion—dynamics. Fundamental to our understanding of mechanics is a knowledge of forces. Forces have a magnitude and a direction and hence are mathematically represented as a vector. Figure 10.1 illustrates a fixed vector force A. It has a magnitude of A newtons or pounds force and is acting at an angle of θ degrees from the horizontal.

Force vectors range from the simplest combinations, all forces acting collinearly as shown in Figure 10.2a, to those acting at the same point or concurrently as in Figure 10.2b, to forces acting coplanarly as in Figure 10.2c. When the forces act on a body, such as a trailer towed by a car (Figure 10.3), the force pulls the trailer and hence is a tension force. As far as the trailer is

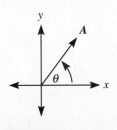

FIGURE 10.1
A fixed vector.

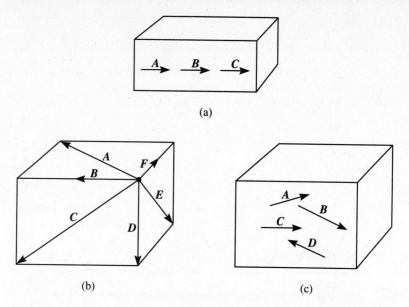

FIGURE 10.2 (a) Collinear forces. (b) Concurrent forces. (c) Coplanar forces.

concerned, it could be pulled by a car or a horse. What matters is the magnitude and direction of the force. Figure 10.4 shows a column, such as in the basement of a house, supporting a weight (force). This force acts in compression on the column.

Vector forces rarely occur at neat angles of 90°, and to simplify analysis we must be able to resolve forces into x- and y-vector components.

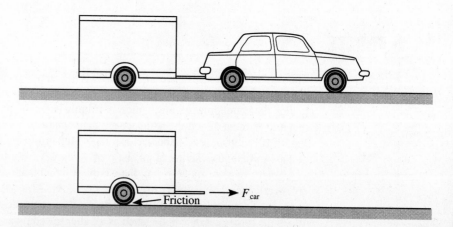

FIGURE 10.3 Horizontal forces acting on a trailer.

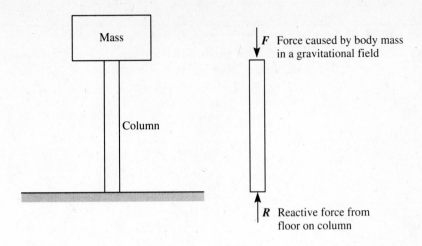

FIGURE 10.4 A compressive force acting on a column.

E X A M P L E
10.1

Resolve the force shown in Figure 10.5 into horizontal and vertical components.

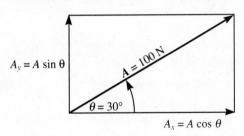

FIGURE 10.5 Resolution of a force A into its rectangular components, A_x and A_y.

SOLUTION:

The force of 100 N acts at an angle of 30° to the horizontal. From trigonometry we know that

$$A_x = A \cos 30 = 100(0.866) = 86.6 \text{ N} \rightarrow$$

and

$$A_y = A \sin 30 = 100(0.5) = 50.0 \text{ N} \uparrow.$$

Very often we determine the resultant of concurrent forces by adding the individual rectangular components of the vector and then determining the resultant vector along the hypotenuse formed by them.

E X A M P L E

10.2

Determine the resultant force (magnitude and direction) from the concurrent force system illustrated in Figure 10.6a.

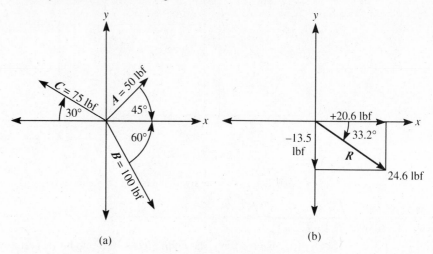

(a)

(b)

FIGURE 10.6 (a) A diagram of the concurrent forces in Example 10.2.
(b) The resultant force.

SOLUTION:

Perhaps the most direct way to solve this problem is to set up a table for the individual components, then sum them.

Force (lbf)	Horizontal component (lbf)	Vertical component (lbf)
100	100 cos 60 = +50.0	−100 sin 60 = −86.6
50	50 cos 45 = +35.6	50 sin 45 = +35.6
75	−75 cos 30 = −65.0	75 sin 30 = +37.5
Total	+20.6	−13.5

Figure 10.6b illustrates the components and the resultant. The resultant R is found from the Pythagorean theorem, with the angle being determined from the definition of either the tan θ, sin θ, and cos θ. Thus,

$$R = (20.6^2 + 13.5^2)^{0.5} = 24.6 \text{ lbf,}$$

and

$$\theta = \tan^{-1} (13.5/20.6) = 33.2°.$$

It is not necessary to resolve two intersecting forces through the summation of components: we may use relationships developed through trigonometric identities and reduce the complexity of the process. Additionally, we will be able

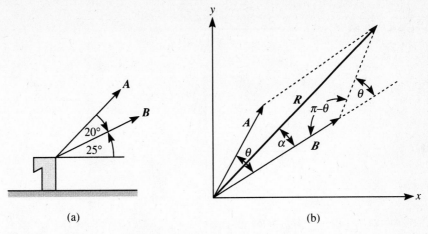

FIGURE 10.7 (a) A tent peg with two concurrent forces acting on it.
(b) The resultant **R** of two forces **A** and **B**.

to determine the resultant and components. Consider the situation in Figure 10.7a, where two forces **A** and **B** are acting on a tent peg. We can determine the resultant using the law of cosines and the law of sines. Figure 10.7b shows the resolution of the forces. Using the general notation of Figure 10.7b, we obtain the magnitudes from the law of cosines:

$$R^2 = A^2 + B^2 - 2AB \ [\cos (\pi - \theta)],$$

and since $\cos (\pi - \theta) = -\cos \theta$, this may be reduced to

$$R^2 = A^2 + B^2 + 2AB \ [\cos \theta]. \qquad 10.1$$

Thus for the angles given in Figure 10.7a, and letting $A = 100$ N and $B = 200$ N, the resultant magnitude is

$$R^2 = (100)^2 + (200)^2 + 2(100)(200) \cos 20° = 87,588$$

$$R = 296.0 \text{ N}.$$

If the resultant magnitude is known and the angles at which the components are acting are also defined, then the magnitudes of components A and B may be determined using the law of the sines:

$$\frac{R}{\sin (\pi - \theta)} = \frac{A}{\sin (\alpha)} = \frac{B}{\sin (\theta - \alpha)}.$$

The magnitudes of the components are

$$A = R \ \frac{\sin \alpha}{\sin (\pi - \theta)} \qquad 10.2a$$

and

$$B = R \ \frac{\sin (\theta - \alpha)}{\sin (\pi - \theta)}. \qquad 10.2b$$

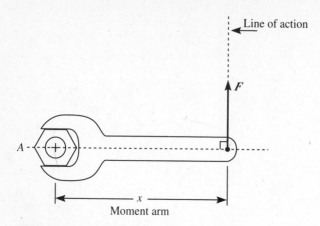

FIGURE 10.8 A moment, $M_A = F \cdot x$, created by a force F acting at a distance x about point A.

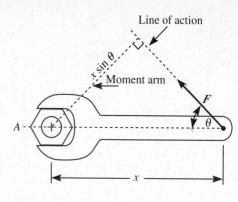

FIGURE 10.9 A moment $M_A = F \cdot x \sin \theta$, created by a force F times the moment arm $x \sin \theta$ about point A.

10.3 MOMENTS

We have thus far considered concurrent forces acting on a body. A *moment* occurs when coplanar forces act on a body at different points, causing an unrestrained body to rotate. A moment about a point is a force times the perpendicular distance to the point. Consider the wrench shown in Figure 10.8, with a force F acting at a distance x. The product of F and x is the *moment about A*, or

$$M_A = F \cdot x, \qquad\qquad 10.3$$

where x is the perpendicular distance from F to A. Moment is a vector with units of newton meters (N·m) or foot-pounds force(ft-lbf). In this case the wrench would rotate in the counterclockwise direction, by convention a positive moment. Clockwise moments are negative. Notice from Figure 10.9 that the distance must be the perpendicular distance from the line of action of the force to some center of rotation, such as A. In this case then, the moment about A is

$$+\curvearrowleft \quad M_A = F \cdot x \sin \theta.$$

■
E X A M P L E
10.3

Determine the moment about C caused by forces A and B in Figure 10.10a.

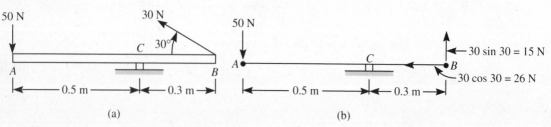

(a)   (b)

FIGURE 10.10 (a) The sketch for Example 10.3. (b) The forces resolved into components.

SOLUTION: The problem may be simplified by noting that force B has horizontal and vertical components. The horizontal component is collinear with pin C and will not cause a moment to occur. The vertical component will cause a moment. Thus, force B may be resolved into its vertical and horizontal components as noted in Figure 10.10b. The moment about C is

$$M_C = (50 \text{ N})(0.5 \text{ m}) + (15 \text{ N})(0.3 \text{ m}) = 29.5 \text{ N·m.}$$

The moments are both positive (counterclockwise). Were one negative and the other positive, they would add algebraically.

◼

Forces that cause moments are often created by an object's weight. Although the weight is not located at a point but distributed over the entire object, it may be considered to act at one point, its center of gravity. For objects with uniform weight distribution the center of gravity corresponds to the geometric center of the object. The symbol ⊗ indicates the center of gravity of an object. Very often a material will have a certain weight, force per linear distance, and can be represented as in Figure 10.11.

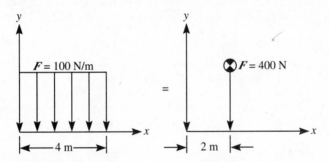

FIGURE 10.11 An object's uniformly distributed force may be replaced by a single force acting at its center of gravity.

10.4 FREE-BODY DIAGRAMS

At this stage it is important to develop a mathematical model of the body that the forces and moments are acting on. This model is called the *free-body diagram,* a schematic drawing of that portion of the body under analysis, isolated from other parts, with the forces and moments from other parts shown. It is "free" in that the other structural elements are represented only as forces and moments. It should be apparent, but may be forgotten when trying to solve a problem, that if the model is wrong, the analysis is useless. Drawing the free-body diagram correctly is an essential first step in mechanics problem analysis.

■
E X A M P L E
10.4

Consider the boom in Figure 10.12a. This could be a cargo boom on a ship; the boom may be adjusted through the pivot at A to position the cargo crate with weight W. Draw the free-body diagram of the boom at point B.

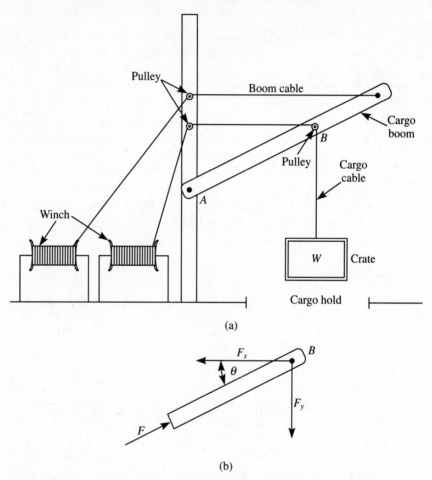

(a)

(b)

FIGURE 10.12 (a) A cargo boom is positioned, raised or lowered, with the boom cable. The cargo is raised or lowered with a separate cargo cable. and (b) its free-body diagram.

SOLUTION:

Figure 10.12b shows the free-body diagram, with only the forces acting on point B. The cable can support forces only in tension, and these forces must act along the axis of the cable. The boom can support forces in tension or compression, with the force vector acting along the boom axis. In this situation the force down, F_y, is equal in magnitude to the weight of the crate, and the horizontal force, F_x, must be equal in magnitude to F_y. The force F in the beam must be the vector sum of these forces, acting in the opposite direction.

■

E X A M P L E
10.5

Draw the free-body diagram for the beam with the loads given in Figure 10.13a.

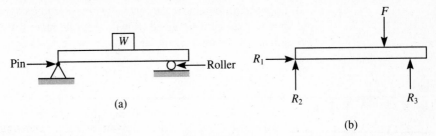

(a)

(b)

FIGURE 10.13 (a) A loaded beam with one hinged end, which supports forces in all directions, and one vertically supported end, which supports only vertical forces. (b) The free-body diagram for the beam.

SOLUTION:

The pin at the left-hand edge of the beam rigidly attaches to the beam, and hence reactive forces in vertical and horizontal directions can occur. The roller support, on the other hand, can transfer only vertical forces. The free-body diagram appears in Figure 10.13b. The forces caused by the support are reactions to the load R on the beam and are denoted by R_1, R_2, and R_3.

Free-body diagrams are not an end in themselves, but are used in conjunction with the conditions of static equilibrium to find unknown forces acting on a body.

10.5 STATIC EQUILIBRIUM

Consider yourself standing stationary on the floor. You are in static equilibrium. There is a force balance. The forces downward, a result of the gravitational acceleration acting on your body mass, are balanced by an equal and opposite force from the ground upward under your feet. This is the weight you feel. It counters the downward force, and a condition of static equilibrium exists. When you fall, static equilibrium is lost until you are firmly on the ground. Because we are three-dimensional beings, the forces balance in each dimension; we could say that the sum of the forces, considering positive and negative directions, is zero for each dimension. Also, there must be no rotation; thus, the sum of the moments about any point on the body must be zero. In this text we consider only two dimensions, x and y, hence

$$\overset{+}{\underset{\rightarrow}{}} \sum_x F_x = 0, \qquad \text{10.4a}$$

$$+\uparrow \sum_y F_y = 0, \qquad \text{10.4b}$$

and

$$+\curvearrowleft \sum M = 0.$$ 10.4c

These three equations we can apply to any problem, and an equal number of unknowns may be determined. Should there be more than three unknowns, the system is indeterminate; the analysis of these systems is beyond our present scope.

■

E X A M P L E

10.6

Figure 10.14a shows a beam with various loads. Determine the reactions at A and B.

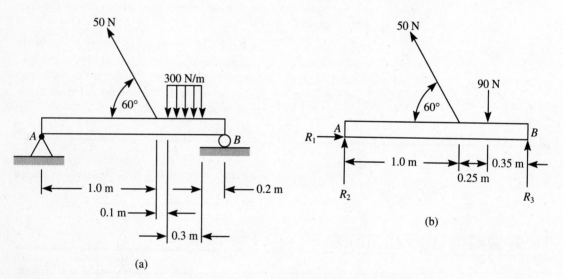

(a)

FIGURE 10.14 (a) A beam loaded per the requirements of Example 10.6. (b) The free-body diagram for the beam.

SOLUTION: The first step is to determine the free-body diagram, as shown in Figure 10.14b. At the fixed-pin end A there are two reactive forces, one vertical (or assumed vertical) and one horizontal to balance the horizontal component of the force in the cable. At the roller end B there is one reactive force upward.

Applying the equilibrium equations, from Equation 10.4a,

$$R_1 - 50 \cos 60 = 0,$$

$$R_1 = 25 \text{ N} \rightarrow.$$

From Equation 10.4b

$$R_2^\uparrow + R_3^\uparrow + 50 \sin 60 - 90 = 0,$$

$$R_2^\uparrow + R_3^\uparrow = 46.7 \text{ N} \uparrow.$$

At this point there are two unknowns, so we need a third independent equation: in equilibrium the sum of moments about any point is zero. We could pick any point, but the wisest choice is either A or B, eliminating either R_2 or R_3, respectively, from Equation 10.4c. We select point A and Equation 10.4c, using the correct sign convention.

$$(1.6)R_3^{\uparrow} - (1.25)90 + (1.0)50 \sin 60 = 0$$

$$R_3 = 43.2 \text{ N} \uparrow$$

$$R_2^{\uparrow} + 43.2 = 46.7 \uparrow$$

$$R_2 = 3.5 \text{ N} \uparrow$$

10.6 STRENGTH OF MATERIALS

Since the beginning of the text when we discussed building a wooden plank bridge, strength of materials has been a recurrent topic. Now that we have a rudimentary understanding of statics and can determine the magnitude of forces acting on an object, we can examine the effect of the forces on the object's material structure.

Stress, rather than force, is used in analyzing a material's strength and is defined as force per unit area. Stress may be a result of forces acting normal to a surface, normal stresses (σ), or acting parallel to a surface, shear stresses (τ). We define these as

$$\sigma = F_n/A \qquad 10.5$$

and

$$\tau = F_t/A, \qquad 10.6$$

where F_n is the normal force, F_t is the tangential force, and A is the cross-sectional area.

Consider the bar shown in Figure 10.15. It is subjected to a force F, and because of the force it lengthens a distance ΔL. The ratio of the change in length to the original length is called the strain ϵ:

$$\epsilon = \Delta L/L. \qquad 10.7$$

In Figure 10.15 the bar is in tension; the force is pulling the material. In compression the force pushes on the bar, trying to contract it. The modulus of elasticity E, called Young's modulus, is the proportionality constant that relates the stress and strain in a material (see Table 10.1). Hook's law, relating stress and strain, was discovered by Robert Hooke in the late 1600s:

$$\sigma = E\epsilon \quad \text{or} \quad E = \sigma/\epsilon. \qquad 10.8$$

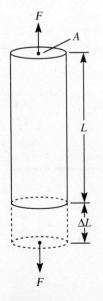

FIGURE 10.15
A bar subjected to a tensile force.

TABLE 10.1 TYPICAL VALUES OF YOUNG'S MODULUS

Material	E (kPa)	E (lbf/in²)
Aluminum	7.0×10^7	1.0×10^7
Brass	9.0×10^7	1.3×10^7
Steel	2.0×10^8	2.9×10^7
Glass	7.0×10^7	1.0×10^7

Equation 10.8 states that stress and strain vary linearly with one another. This is valid for the elastic region of a stress-strain diagram, as in Figure 10.16. If the force is released in the linear (elastic) region, the material will return to its original state. Once the yield point is reached, however, the material does not behave elastically, and a permanent deformation occurs. If stretched far enough, the material will break.

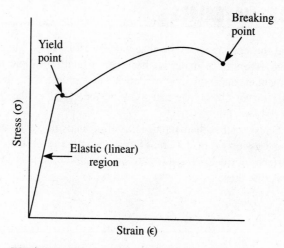

FIGURE 10.16 A stress-strain diagram.

E X A M P L E
10.7

Figure 10.17a shows a steel bar machined to two different diameters and subjected to a load of 1000 kN. Determine the stress in each section and the total extension of the bar under load.

SOLUTION:

From Equation 10.5 we can determine the stress.

$$\sigma_{0.2} = F_n/A = \frac{1000 \text{ kN}}{\pi/4 \ (0.2)^2 \ \text{m}^2} = 31 \ 830 \text{ kPa}$$

$$\sigma_{0.1} = F_n/A = \frac{1000 \text{ kN}}{\pi/4 \ (0.1)^2 \ \text{m}^2} = 127 \ 320 \text{ kPa}$$

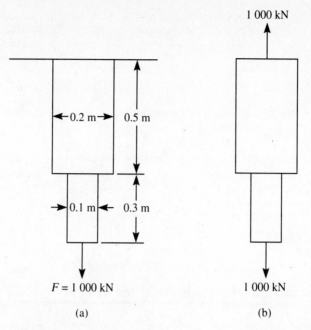

FIGURE 10.17 (a) An axially loaded bar of two different diameters. (b) The free-body diagram for the bar.

From Equation 10.8 we can determine the strain and from that the length increase for each diameter from Equation 10.7.

$$\epsilon = \sigma/E$$

$$\epsilon_{0.2} = (31\ 830)/(2 \times 10^8) = 1.6 \times 10^{-4}$$

$$\epsilon_{0.1} = (127\ 320)/(2 \times 10^8) = 6.4 \times 10^{-4}$$

$$\Delta L = \epsilon L$$

$$\Delta L_{0.2} = (0.5)(1.6 \times 10^{-4})(1000\ \text{mm/m}) = 0.08\ \text{mm}$$

$$\Delta L_{0.1} = (0.3)(6.4 \times 10^{-4})(1000\ \text{mm/m}) = 0.19\ \text{mm}$$

Thus, the total length increase is 0.27 mm, not much at all.

■

    The analysis of materials in tension or compression is similar; the material is acted upon by axial forces. In shear, however, the force is perpendicular to the material's axis. In Figure 10.18a the force acts perpendicular to the cross-sectional area of the rivet. Figure 10.18b shows a rivet in double shear; the shear area is twice the cross-sectional area of the rivet.

    When you are designing an object, you must assume a safe allowable value for the stresses and then calculate the actual values, assuring that the actual values are less than the working value. The normal and shear stress values for

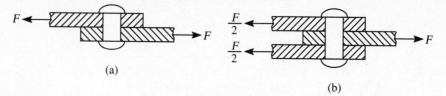

FIGURE 10.18 A rivet being subjected to (a) single shear and (b) double shear.

materials are known at the yield point, σ_y and τ_y, respectively. Dividing these values by the desired factor of safety, S, yields the allowable value of stress that is used in your sizing calculations (see Table 10.2). Thus,

$$\sigma_{allowable} = \sigma_y/S,$$ 10.9

and

$$\tau_{allowable} = \tau_y/S.$$ 10.10

TABLE 10.2 YIELD STRESS VALUES FOR METALS

Material	Normal Yield Stress		Shear Yield Stress	
	10^3 lbf/in^2	MPa	10^3 lbf/in^2	MPa
Aluminum	50	345	30	207
Bronze	20	138	11	76
Steel	90	620	52	358

In designing a structural element from aluminum with a factor of safety of 1.75, the allowable normal stress is 197 MPa and the allowable shear stress is 118 MPa.

10.7 MATERIALS SCIENCE

Engineers not only analyze the forces acting on components and devices, they also select the material that the object is made from. In mechanical engineering applications one tends to think of metals when considering material selection. Older curricula often included a course in metallurgy; this course has been replaced by a more general course in materials science. The variety of materials has expanded tremendously in the past 25 years, with seemingly limitless types of materials available. Five general categories embrace the variety of materials practicing engineers will encounter: metals, ceramics, plastics, semiconductors, and composites.

The reason that metals dominated materials selection can be understood by glancing at the periodic table in Figure 10.19. The shaded elements are classed as metals because of their atomic bonding characteristics. Metals share electrons

1A																	O
1 H	IIA											IIIA	IVA	VA	VIA	VIIA	2 He
3 Li	4 Be											5 B	6 C	7 N	8 O	9 F	10 Ne
11 Na	12 Mg	IIIB	IVB	VB	VIB	VIIB	VIII			IB	IIB	13 Al	14 Si	15 P	16 S	17 Cl	18 Ar
19 K	20 Ca	21 Sc	22 Ti	23 V	24 Cr	25 Mn	26 Fe	27 Co	28 Ni	29 Cu	30 Zn	31 Ga	32 Ge	33 As	34 Se	35 Br	36 Kr
37 Rb	38 Sr	39 Y	40 Zr	41 Nb	42 Mo	43 Tc	44 Ru	45 Rh	46 Pd	47 Ag	48 Cd	49 In	50 Sn	51 Sb	52 Te	53 I	54 Xe
55 Cs	56 Ba	57 La	72 Hf	73 Ta	74 W	75 Re	76 Os	77 Ir	78 Pt	79 Au	80 Hg	81 Tl	82 Pb	83 Bi	84 Po	85 At	86 Rn
87 Fr	88 Ra	89 Ac															

58 Ce	59 Pr	60 Nd	61 Pm	62 Sm	63 Eu	64 Gd	65 Tb	66 Dy	67 Ho	68 Er	69 Tm	70 Yb	71 Lu
90 Th	91 Pa	92 U	93 Np	94 Pu	95 Am	96 Cm	97 Bk	98 Cf	99 Es	100 Fm	101 Md	102 No	103 Lw

FIGURE 10.19 The periodic table. Elements shaded are inherently metallic in nature.

between atoms in a nondirectional bond that manifests high electrical conductivity. In addition, other properties such as ductility are associated with metals. Metals may be formed into a variety of shapes, from steel girders to sinks. Many of the metallic elements can combine to form alloys, metals such as brass (copper and zinc) and steel (iron and carbon).

Ceramic materials, including glass, are metallic oxides, such as silicon dioxide (SiO_2) and aluminum trioxide (Al_2O_3). Again, the atomic bonding of this material establishes its identity. These materials have not only an ionic bonding but also a strong covalent bond between the atoms. The covalent bond is electron sharing that is highly directional and results in low values of electrical conductivity. Whereas metals are malleable on the macroscopic level, ceramics are brittle, exhibiting low ductility and malleability. Ceramics can withstand extremely high temperatures without loss of their strength properties, even in chemically active environments such as combustion spaces. Many automotive engine parts are ceramic or ceramic-coated, and research is currently underway to develop a ceramic engine. This engine would not need a radiator, and hence much of the energy now lost as heat could be converted to work, increasing the engine's efficiency.

Perhaps the group of materials most closely associated with the recent industrial era is plastics, more fundamentally called polymers. A polymer is produced by combining many hydrocarbon molecules together in a long chain. The name comes from *poly*, many, and *mer*, a single hydrocarbon molecule. Thus, polyethylene is a polymer formed by combining many ethylene molecules (C_2H_4) together. The atomic bonding of polymers is characterized by strong covalent bonds within a chain and weak bonds between chains. The weaker bonds between the chains allow for plastics being highly ductile on one hand, but having low strength and melting point characteristics on the other. Polymers do not result strictly from joining hydrocarbons together but may include other substances such as nitrogen in the molecular chain. Nylon is just such an instance.

IA							VIII										O
1 H	IIA									IB	IIB	IIIA	IVA	VA	VIA	VIIA	2 He
3 Li	4 Be											5 B	6 C	7 N	8 O	9 F	10 Ne
11 Na	12 Mg	IIIB	IVB	VB	VIB	VIIB				IB	IIB	13 Al	14 Si	15 P	16 S	17 Cl	18 Ar
19 K	20 Ca	21 Sc	22 Ti	23 V	24 Cr	25 Mn	26 Fe	27 Co	28 Ni	29 Cu	30 Zn	31 Ga	32 Ge	33 As	34 Se	35 Br	36 Kr
37 Rb	38 Sr	39 Y	40 Zr	41 Nb	42 Mo	43 Tc	44 Ru	45 Rh	46 Pd	47 Ag	48 Cd	49 In	50 Sn	51 Sb	52 Te	53 I	54 Xe
55 Cs	56 Ba	57 La	72 Hf	73 Ta	74 W	75 Re	76 Os	77 Ir	78 Pt	79 Au	80 Hg	81 Tl	82 Pb	83 Bi	84 Po	85 At	86 Rn
87 Fr	88 Ra	89 Ac															

58 Ce	59 Pr	60 Nd	61 Pm	62 Sm	63 Eu	64 Gd	65 Tb	66 Dy	67 Ho	68 Er	69 Tm	70 Yb	71 Lu
90 Th	91 Pa	92 U	93 Np	94 Pu	95 Am	96 Cm	97 Bk	98 Cf	99 Es	100 Fm	101 Md	102 No	103 Lw

FIGURE 10.20 The periodic table. Naturally occurring elemental semiconductors are darkly shaded, and semiconducting compound–forming elements are lightly shaded.

The current industrial era, or perhaps the computer era, is characterized by the semiconductor materials. Solid-state electronics relies on semiconductors, materials that neither insulate nor conduct well. Semiconductors cannot be readily categorized by atomic bonding, though covalent bonding predominates. Figure 10.20 shows the periodic table with those elements that form semiconductors shaded. In column IV A three elements, silicon (Si), germanium (Ge), and tin (Sn), are naturally occurring semiconductors. More common are semiconductor compounds formed by combining an element from the left of column IV A with one to the right of the column. Commonly found semiconductor compounds are gallium-arsenic (GaAs), indium-antimony (InSb), and cadmium-sulfide (CdS).

Magnification of a fibrous material. (Courtesy of ASME)

A final group of materials are composites. Normally we think of composites as being modern materials, such as the graphite epoxy widely used in the aerospace industry. However, more mundane composites are concrete and wood. Composites combine elements, metals, ceramics, and polymers to form a new material. Very often the combining of the materials will produce a new material with properties superior to those of the constituents. For instance, fiberglass is formed by combining glass fibers with a polymer. The resulting composite has excellent strength characteristics, due to the glass fibers, and very good malleability, due to the polymer.

10.8 DYNAMICS

When there is a force imbalance on a body, that is, $\Sigma F_x / 0$, motion occurs. Dynamics is the study of rigid body motion, as opposed to fluid motion, and is of great interest to mechanical engineers, who often design moving machinery parts. As the parts of the body will be in motion, a knowledge of velocity and acceleration is required. The study of dynamics becomes complex rather quickly, because most movement requires the analysis of a body moving in two or three dimensions. Some interesting introductory aspects are well within our abilities, however, and we can gain some understanding of body motion from these.

The equations describing motion in one dimension are readily developed. Assume the direction of motion to be horizontal for the time being, and let the x-axis represent distance, or position. Position is the location of an object at a certain value of x. The initial position is x_1, and the time associated with the object's location at x_1 is t_1. Velocity v is the vector defined as the change of position with time, m/s or ft/sec:

$$v = \frac{x_2 - x_1}{t_2 - t_1} = \frac{\gamma \Delta x}{\Delta t} \text{ m/s (ft/sec)}. \qquad 10.11$$

If the velocity changes with time, that is called acceleration $\vec{a}$, m/s^2 or ft/sec^2:

$$a = \frac{v_2 - v_1}{t_2 - t_1} = \frac{\Delta v}{\Delta t}. \qquad 10.12$$

Equations 10.11 and 10.12 can be used to derive other, more useful relationships. In this section we will consider acceleration that is constant with time, such as Earth's gravitational acceleration. At a given location it remains a constant value; 9.8 m/s^2 or 32.174 ft/sec^2 is the standard value. At $t = 0$ the accelerating object has some initial velocity and is located at a given distance on the x-axis. These *boundary conditions* to the problem are written as

$$t = 0; \ x = x_0; \ v = v_0; \ a = \text{constant};$$

and

$$t = t; \ x = x; \ v = v; \ a = \text{constant}.$$

If we substitute these into Equation 10.12, we obtain

$$v = v_0 + at. \qquad 10.13$$

Thus the velocity at any time t is equal to the initial velocity at $t = 0$ plus the acceleration multiplied by the time. This is also the equation of a straight line with slope a, intercept v_0. Solving Equation 10.11 for the change in horizontal distance while dropping the vector notation gives

$$x - x_0 = v\Delta t, \qquad 10.14$$

where v is the average velocity that occurs during the time interval Δt. Using the boundary conditions we imposed earlier,

$$v = \frac{v_0 + v}{2}. \qquad 10.15$$

Substituting into Equation 10.14 yields

$$x - x_0 = \tfrac{1}{2}(v + v_0)t. \qquad 10.16$$

Equation 10.16 relates distance, velocity, and time. If we wish to know the distance in terms of the acceleration and time, we can use Equation 10.13 for the velocity and substitute this into Equation 10.16.

$$x - x_0 = \tfrac{1}{2}(v_0 + at + v_0)t$$

$$x = x_0 + v_0 t + \tfrac{1}{2}(at^2) \qquad 10.17$$

Lastly, we may also need an equation that determines velocity in terms of acceleration and distance. Solving Equation 10.13 for the time, $t = (v - v_0)/a$, and substituting into Equation 10.16,

$$x - x_0 = \tfrac{1}{2}(v + v_0)\left(\frac{v - v_0}{a}\right) = \frac{v^2 - v_0^2}{2a};$$

$$v^2 = v_0^2 + 2a(x - x_0). \qquad 10.18$$

Now we will apply these equations.

E X A M P L E
10.8

During the construction of a skyscraper, a worker drops a rivet from the top of a large building. What is its velocity after 4 seconds? If the building is 150 meters tall, what is the velocity of the rivet upon impact with the ground?

SOLUTION:

Assume the acceleration is 9.8 m/s^2. The situation you are analyzing is that of an object acted upon by constant acceleration with a boundary condition of $x = 0$, $v = 0$ at $t = 0$. From Equation 10.13 we can determine the velocity after 4 seconds.

$$v = 0 + (9.8 \text{ m/s}^2)(4 \text{ s}) = 39.2 \text{ m/s}$$

Using Equation 10.18 allows us to determine the final velocity.

$$v^2 = 0 + 2(9.8 \text{ m/s}^2)(150 \text{ m}) = 2940 \text{ m}^2/\text{s}^2$$

$$v = 54.2 \text{ m/s}$$

The time may be determined from Equation 10.13.

$$t = (54.2 - 0 \text{ m/s})/(9.8 \text{ m/s}^2) = 5.5 \text{ s}$$

■

E X A M P L E
10.9

A dragster has a velocity of 120 miles/hour at the end of a quarter-mile track and must decelerate uniformly and come to a stop in 2.5 seconds. What is the deceleration? How far does the dragster travel before stopping?

SOLUTION:

In this situation the acceleration is negative, a deceleration. The object, the dragster, has a positive initial velocity at time zero. Furthermore, we must change the units from miles per hour to feet per second. From Appendix 2, $(120 \text{ mi/h})(1.467) = 176.0 \text{ ft/sec}$.

The boundary conditions are $t = 0$, $v_0 = 176.0 \text{ ft/sec}$, $x = 0$ and $t = 2.5 \text{ sec}$, $v = 0$, $x = \text{L}$.

From Equation 10.12 we can determine the average deceleration to be

$$a = \frac{0 - 176}{2.5 - 0} = -70.4 \text{ ft/sec}^2.$$

The value is negative, indicating deceleration. Solving Equation 10.16 for the distance yields

$$x = L = 0 + 0.5(0 + 176 \text{ ft/sec})(2.5 \text{ sec}) = 220 \text{ ft}.$$

■

10.9 STATIC FRICTION

Forces of friction result from the resistance to motion between a body and a surface. These frictional forces are essential to our daily lives, walking and running, or driving a car. The two types of friction are static, with no body motion, and kinetic, with body motion across a surface.

Investigation of frictional forces has shown that both static and kinetic frictions are proportional to the normal force acting on a surface. Consider the block in Figure 10.21a. The normal force N is equal and opposite to the force due to weight W. Now we wish to push the block and exert a force F_x, as illustrated

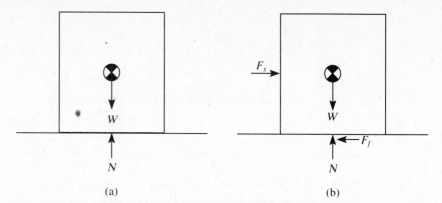

FIGURE 10.21 (a) The normal force acting vertically upward in opposition to a container's weight. (b) The frictional force F_f opposes the container's movement, caused by force F_x, and is a function of the normal force N.

in Figure 10.21b. The resistance to the motion is a frictional force F_f, parallel to the surface of contact. The maximum value of the frictional force is

$$F_f = \mu_s N \qquad\qquad 10.19$$

where μ_s is the coefficient of static friction. When the horizontal force that you exert is greater than the resisting frictional force, the block starts to move. At the onset of movement the coefficient of friction decreases, and the kinetic coefficient of friction, μ_k, is used. The equation for the kinetic friction is the same as Equation 10.19. Table 10.3 lists the typical values for the coefficients of friction, static and kinetic. Notice that on average the kinetic coefficient is about 25 percent less than the static value. In the simplest view this is caused by a decrease in the relative roughness between the surfaces.

TABLE 10.3 APPROXIMATE VALUES FOR COEFFICIENTS OF FRICTION

Surfaces	Static coefficient (μ_s)	Kinetic coefficient (μ_k)
Steel on steel	0.74	0.57
Aluminum on steel	0.61	0.47
Rubber on concrete	1.0	0.8
Wood on wood	0.25–0.5	0.2
Glass on glass	0.94	0.4
Metal on metal (lubricated)	0.15	0.06

■
E X A M P L E
10.10

A box sits on a ramp leading to a floating dock. As the tide goes out, the water level decreases and the angle of the incline increases. If the coefficient of static friction is estimated to be 0.5, what is the maximum angle of the ramp before the box starts sliding?

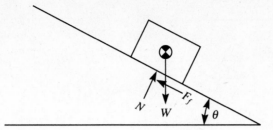

FIGURE 10.22 A free-body diagram for Example 10.10.

SOLUTION: Figure 10.22 is the free-body diagram for the box and ramp. The friction force opposes the force due to the component of the box's weight parallel to the plane's surface.

$$N = W \cos \theta$$

$$F_f = \mu_s N = \mu_s W \cos \theta$$

$$F_{parallel} = W \sin \theta$$

$$W \sin \theta = \mu_s W \cos \theta$$

$$\theta = \tan^{-1} 0.5 = 26.5°$$

Thus, when the incline becomes greater than 26.5°, the box will start to slide.

■

10.10 KINETIC FRICTION

At this point the force on the block on the inclined plane is great enough to overcome friction, causing the box to slide. In Figure 10.23 a block of mass m is released from rest on an incline a distance L from the base. The angle of the incline to the horizontal is θ degrees. Newton's second law states that

> A mass m acted upon by an external force will have an acceleration in the direction of the force and directly proportional to the force.

Mathematically this is

$$F = ma \qquad\qquad 10.20$$

or where there is a net force resulting from the interaction of several forces,

$$F_{net} = ma.$$

Considering the x dimension,

$$F_x = ma_x = ma. \qquad\qquad 10.21a$$

and in the y direction,

$$F_y = W = mg \qquad\qquad 10.21b$$

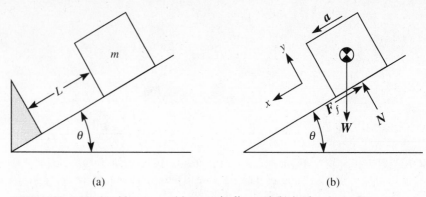

FIGURE 10.23 A block with mass m (a) on an incline and (b) its free-body diagram.

where g is the acceleration due to gravity. The sum of the forces is the algebraic sum of the force due to acceleration acting down the incline and the frictional force opposing it. If the coefficient of friction is the kinetic one and the downward force is great enough to overcome the static friction, we substitute into Equation 10.21.

$$W \sin \theta - F_f = ma$$

$$mg \sin \theta - \mu_k mg \cos \theta = ma$$

$$g(\sin \theta - \mu_k \cos \theta) = a = \text{constant} \qquad 10.22$$

Thus, we have a situation where the acceleration is constant and in one direction. Equations 10.11 to 10.18 apply.

E X A M P L E
10.11

Let the block in Figure 10.23a have a mass of 50 kg, $L = 20$ m, $\mu_k = 0.2$, and $\theta = 30°$. Determine the time to reach the base and the velocity at that time.

SOLUTION:

We check to see if the box will slide. From the section on static friction we noted that if $\tan \theta > \mu_s$, the box will slide. $\tan 30 = 0.577$, which is greater than 0.2, so the box will slide. Since the box will slide, calculate the acceleration from Equation 10.22:

$$a = (9.8 \text{ m/s}^2)(\sin 30 - 0.2 \cos 30) = 3.2 \text{ m/s}^2.$$

The time to reach the base is found using Equation 10.17.

$$20 = 0 + 0 + (0.5)(3.2)t^2$$

$$t = 3.54 \text{ s}$$

The velocity at this time may be found from several equations. Using Equation 10.13,

$$v = 0 + (3.2)(3.54) = 11.3 \text{ m/s}.$$

10.11 TRUSSES AND FRAMES

We know that bridges and buildings are more complex structures than those we considered when analyzing a single beam subjected to loads in Section 10.4. Figure 10.24a shows a simple bridge. The key to analyzing this bridge design, which is probably quite familiar, is knowing how trusses transmit loads. Notice that the load carried by the floor is transmitted to the stringers, to the crossbeams, and finally to the side trusses.

The method of joints directly determines the forces in the truss members, though it requires some assumptions:

1. The trusses are weightless and straight. This in essence means that the weight of the loads is much greater than the weight of the truss and that the beams composing the truss do not bend, they remain rigid.
2. All the joints are pinned, not fixed. This allows the junction to pivot, again eliminating bending.

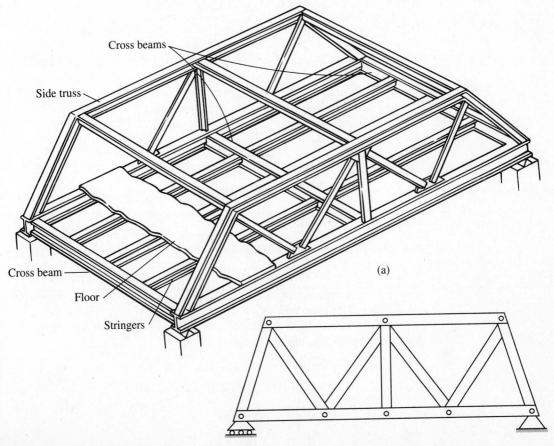

(a)

(b)

FIGURE 10.24 (a) A simple truss bridge.
(b) The side view.

3. All external loads act at the joints of the truss. From the sketch of the bridge, we see that the loads in reality are transmitted in large part by the crossbeams to the joints.

Because the loads are assumed to be transmitted by the static members to the joints, we can model the bridge by viewing a truss with loads applied to it. Figure 10.24b shows the truss from the bridge. Considering that each member between the pins contains a separate force and that there are two vertical reactive forces and one horizontal reactive force, a total of sixteen unknowns must be determined to solve for the forces in Figure 10.24b. For static equilibrium we know that the sum of the forces in the x and y directions is zero; for eight pins, 16 equations must be solved. This is a fairly lengthy procedure, so let's illustrate the method with a simpler triangular truss.

■

E X A M P L E
10.12

SOLUTION:

Determine the forces F_1, F_2, and F_3 and the reactions R_1, R_2, and R_3 for the triangular truss illustrated in Figure 10.25a.

The static members AB, BC, and AC will be in either tension or compression. In solving for the forces we will assume a given direction, say tension; if the sign of the answer is negative, it means we assumed the wrong direction. The magnitude of the term will be correct, however. Before we can write the equations, we draw a free-body diagram for the pins, Figure 10.25b.

At pin B the conditions of static equilibrium yield

$$\overset{+}{\rightarrow} \sum F_x = 0 = -F_1 \cos \alpha + F_2 \cos \alpha$$
$$F_1 = F_2$$

and

$$+\uparrow \sum F_y = 0 = -1000 - F_1 \sin \alpha - F_2 \sin \alpha.$$

Since $F_1 = F_2$ this yields

$$2F_1 \sin \alpha = -1000$$

$$F_1 = -833.3 \text{ N} = F_2.$$

Note the sign is negative, denoting that the forces F_1 and F_2 are compressive, not tensile, as our intuition tells us.

At pin A, assuming F_1 is acting as a compressive force as we determined from pin B, static equilibrium yields

$$\overset{+}{\rightarrow} \sum F_x = 0 = -F_1 \cos \alpha + F_3$$
$$F_3 = (833.33)(4/5) = 666.67 \text{ N}$$
$$+\uparrow \sum F_y = R_1 - F_1 \sin \alpha$$

and

$$R_1 = (833.33)(3/5) = 500 \text{ N} \uparrow$$

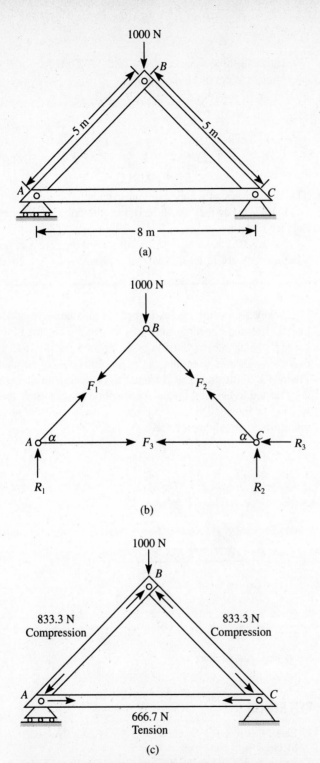

FIGURE 10.25 (a) A triangular truss. (b) A free-body diagram of the truss, with loads acting at the pins. (c) The internal forces.

We know all three forces, F_1, F_2, and F_3. To solve for R_2 and R_3 consider the conditions for equilibrium at pin C.

$$\xrightarrow{+} \sum F_x = 0 = F_3 + F_2 \cos \alpha - R_3$$

$$R_3 = -666.67 + (833.33)(4/5) = 0$$

$$+ \uparrow \sum F_y = 0 = R_2 - F_2 \sin \alpha$$

$$R_2 = 500 \text{ N} \uparrow$$

This also serves as a check on our calculations; we know that R_1 plus R_2 must equal 1000 N, as they do. In addition, the horizontal components of F_1 and F_2 balance one another.

Figure 10.25c shows the magnitudes of the forces in the truss members and whether they are in tension or compression.

∎

A key feature of trusses is that the members are not subjected to bending loads. Often one or more members are rigidly attached together, however, not pinned, which means the forces do not act parallel and perpendicular to the beam's major axis. Bending can occur. This type of structure is called a *frame*. Figure 10.26 illustrates an A-frame. Even though a roller is shown, the structure is subjected to bending forces. The method of solution is more complicated (and will not be included) as equilibrium conditions consider not only force balances, but moment equations are well.

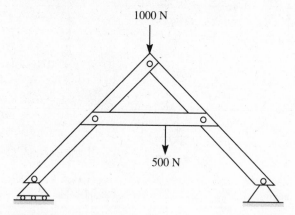

FIGURE 10.26 An A-frame.

REFERENCES

1. Beer, F.P., and Johnson, E.R., Jr. *Vector Mechanics for Engineers*. 5th ed. McGraw-Hill, New York, 1988.
2. Popov, E.P. *Engineering Mechanics of Solids*. Prentice-Hall, Englewood Cliffs, N.J., 1990.
3. Shackelford, J.F. *Introduction to Materials Science for Engineers*. Macmillan, New York, 1985.

PROBLEMS

10.1 In Figure 10.6a let $A = 100$ N, $B = 80$ N, and $C = 50$ N. Determine the resultant force's magnitude and direction.

10.2 It is possible to resolve forces graphically by drawing the magnitudes and angles to scale and joining the forces tip to tail. Perform this for the values in Example 10.2.

10.3 Determine the resultant force (magnitude and direction) for the following concurrent force systems.

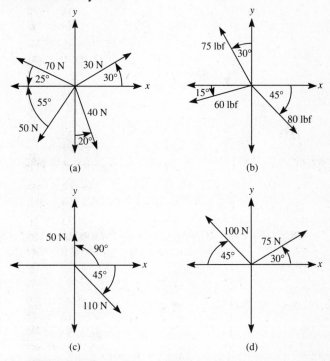

10.4 Given that the resultant force acting on the tent peg shown in Figure 10.7a is 100 lbf and the angles are as indicated, determine the values of A and B.

10.5 Two tugboats are required to pull a barge as illustrated in the following sketch. The horizontal force acting at C is 20,000 N. Determine the force (tension) in the ropes (actually called lines in nautical jargon) AC and BC.

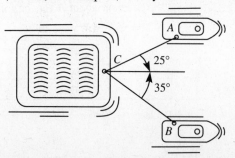

10.6   Determine the moment (torque) in foot-pounds force about point A for the wrench.

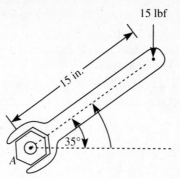

15 lbf

15 in.

35°

A

10.7   Determine the net moments about A and B for the beam.

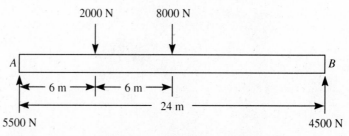

2000 N   8000 N

A B

6 m   6 m

24 m

5500 N   4500 N

10.8   This wheelbarrow must be supported at an angle of 20° to the horizontal. Determine the force required. (Hint: Sum moments about A.)

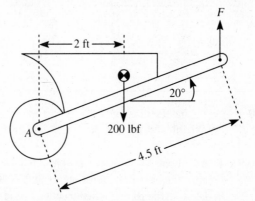

F

2 ft

20°

A 200 lbf

4.5 ft

10.9   Determine the moment about A for the following sketch.

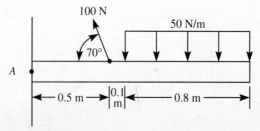

100 N

50 N/m

70°

A

0.5 m   0.1 m   0.8 m

10.10 Find the components of the box's weight parallel and perpendicular to the inclined plane.

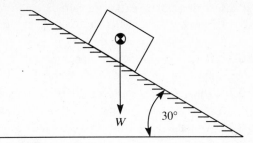

10.11 A horizontal beam is 15 m long and weighs 2000 N. It has pinned supports at the extreme left end and 3 m from the right end. In addition a concentrated load of 500 N is 5 m from the left end. Determine the reactions at the supports.

10.12 Refer to Figure 10.12a. The crate contains a 2500-pound car, and the angle is 40°. Determine the tension in the cable and the x and y components of the force at B.

10.13 Refer to Figure 10.14a. Let the angle be 120°, and determine the reactions at A and B.

10.14 A 15-ft plank that weighs 7 lb/ft is horizontal, attached rigidly to the wall at the right end and supported by a vertical cable 3 ft from the left end. A person weighing 150 lbf is standing in the center of the plank. Determine the reaction, magnitude, and direction at both supports.

10.15 The person in Problem 10.14 now moves past the cable and is standing 1.5 ft from the left-hand side. Determine the reactions, magnitude, and direction at both supports.

10.16 A tire has a diameter of 0.65 m and supports a vertical load of 500 kg. A force, acting at the centerline, must be sufficient to cause the tire to move over a 10-cm curb. What is the amount of force?

10.17 Refer to Figure 10.15. Let the force be 1000 lbf, the length be 2 ft, and the diameter be 1 in. Calculate the normal stress and strain for steel, aluminum, and brass.

10.18 Refer to Figure 10.13. Let the bar be hollow with an inside diameter of 10 cm, an outside diameter of 12.5 cm, and a length of 1 m. The bar is subjected to a loading of 2000 kN. Determine the elongation and the normal stress if the material is aluminum.

10.19 Refer to Figure 10.17a. Let the bar be bored to an inside diameter of 5 cm. Determine the loadings per Example 10.7.

10.20 Two 1-cm diameter rivets join two metal sheets together. If the force pulling the sheets is 35 kN, determine the average shear stress in each rivet.

10.21 Two plastic parts are butted together and glued. The parts' mating surfaces have dimensions of 100 mm by 5 mm. If a tensile force of 5000 N is applied, what is the average normal stress at the interface?

10.22 A riveted connection must support a load of 2000 lbf in single shear. What diameter steel rivet should be used if the factor of safety is 1.5?

10.23 A riveted connection supports a load of 1500 kN in double shear, per Figure 10.18a, with two rivets. What is the rivet diameter if the rivets are steel and the factor of safety is 1.5?

10.24 Determine for the load conditions in Problem 10.12 the size of the steel pin at point B. The pin is made of steel, and the factor of safety is 2.

10.25 The cable in Problem 10.14 has an allowable stress of 8500 psi. Determine its diameter for the worst load condition.

10.26 A product is assembled while on a conveyor belt that is 15 m long and moves at 3 cm/s. Determine the time to complete a product assuming it takes the full 15 m to complete it.

10.27 In parts of New York City there are on average 20 blocks per mile, with stoplights at every other intersection. The light changes are staggered by 30 s from one to the next. Determine the constant speed you should drive for uninterrupted travel.

10.28 An elevator's steady velocity is 5 m/s. It achieves this in 2.5 s. What is the elevator's average acceleration?

10.29 An elevator's mass is 1000 kg. It accelerates at 1 m/s^2. Determine the force in the cable pulling the elevator.

10.30 The elevator in Problem 10.29 carries ten people with an average mass of 70 kg each. If the acceleration remains the same, what is the force in the cable?

10.31 A ball with a mass of 0.25 kg is thrown vertically upward with an initial velocity of 13 m/s. What is the maximum height it will reach?

10.32 What will the velocity of the ball in Problem 10.31 be when it has traveled 3 m?

10.33 A ball is thrown straight upward from the top of a building, with an initial velocity of 20 m/s. The ball misses the building on its downward vertical path. The building is 100 m tall. Determine the time for the ball to reach its maximum height, what the maximum height is, and the time for the ball to hit the ground.

10.34 An object has a velocity of 12 m/s when its distance is 3 m. Two seconds later the distance is 11 m. What is the acceleration of the object?

10.35 The Moon's gravitational acceleration is one-sixth that of Earth's. An astronaut throws a lunar rock upward with a velocity of 20 m/s. What height will the rock attain, and how long will it stay in motion?

10.36 You are driving a car at 120 km/h when you see an accident ahead. Stepping on the brakes, you decelerate the car at -25 m/s^2. How long does it take you to stop, and what distance does the car travel?

10.37 You are designing a target backstop on a rifle range. You determine that a thick wooden backstop is desirable to prevent ricochet. To determine how thick the wood should be, you set up an experiment and collect the following data: A bullet traveling at a velocity of 400 m/s decelerates to

300 m/s while passing through a 4-cm-thick board. Determine the deceleration. Based on this value, determine the minimum thickness required for the backstop.

10.38 When you push a box up an incline, the direction of the pushing force is not parallel to the surface, and a portion of it increases the normal force and hence the frictional force. The coefficient of static friction is 0.30. Determine for the values given in the figure whether or not the block will move.

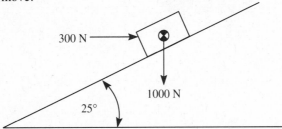

10.39 If the force in Problem 10.38 doubles, will the block move?

10.40 Write a computer program for the sliding-block problem on an inclined plane. The length, angle, and coefficient of friction are the variables. Run the program for angles of 15°, 30°, and 45° and coefficients of friction of 0.05, 0.1, and 0.8 for a length 20 m. The program should determine if the block will slide or not, the time to reach the bottom, and the velocity when it reaches the bottom.

10.41 Refer to Figure 10.25a. Let the force at B be horizontal, acting in the positive x direction and having a magnitude of 1000 N. Compute the forces in each of the members.

10.42 Refer to Figure 10.25a. Let there be a 1000-N horizontal force acting in the negative x direction and a 1000-N force acting vertically downward at B. Compute the forces in each of the members.

10.43 For the truss below, compute the forces in each of the members.

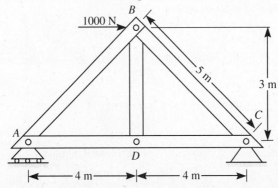

10.44 Refer to the figure in Problem 10.43. Let an additional force act vertically downward at B with a value of 500 N. Compute the forces in each of the members.

11

ENERGY SYSTEMS

■ To apply conservation of mass equations.

■ To define the types of energy forms used in thermodynamic analysis.

■ To apply the first law of thermodynamics to a variety of situations.

■ To explore the implications of the second law of thermodynamics.

■ To examine energy utilization and future energy problems.

(Photo courtesy of Standard Oil Company.)

*M*any studies indicate that in the near future the world will have greatly diminished petroleum and mineral supplies, resources that sustain the standard of living the industrialized world enjoys. All aspects of manufacture and living will have to become more energy efficient as the costs and scarcity of resources increase. Engineers must create new systems and devices that will reduce our energy consumption and improve our energy utilization.

11.1 INTRODUCTION

This chapter has two focuses, the first being an understanding of fundamental system energy analysis from a thermodynamics viewpoint. Once the rudimentary thermodynamic skills are developed, we can apply these to a variety of energy systems that have major import and impact on our society, for example, the energy use in agriculture, electric power generation, and transportation systems.

Thermodynamics is the science that is devoted to understanding energy in all its forms (mechanical, electrical, chemical) and how energy changes forms: the transformation of chemical energy into thermal energy, for instance. *Thermodynamics* is derived from the Greek words *therme,* meaning heat, and *dynamis,* meaning strength, particularly applied to motion. Literally, then, thermodynamics would mean ''heat strength,'' implying such things as the heat liberated by the burning of wood, coal, or oil. Actually, if the word *energy* is substituted for *heat,* one can come to grips with the meaning and scope of thermodynamics. It is the science that deals with energy transformations: the conversion of heat into work, or of chemical energy into electrical energy. Both of these are energy transformations, and thermodynamics is the science that provides the tools to analyze them. The power of thermodynamics lies in its usefulness in analyzing a wide range of energy systems using only a few tenets. Two primary ones are the first and second laws of thermodynamics.

11.2 CONSERVATION OF MASS

The expression for mass flow rate is frequently used in conjunction with thermodynamic analysis. Let the mass flow rate in kg/s or lbm/sec be denoted as $\dot{m}$. The conservation law states that for steady-state conditions the mass flow entering a device must equal the mass flow leaving the device. Thus,

$$\dot{m}_{in} = \dot{m}_{out}.$$

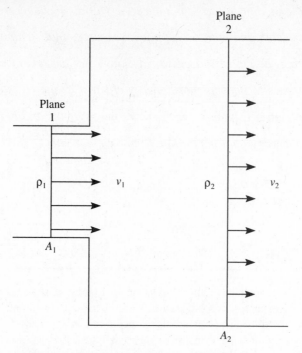

FIGURE 11.1 One-dimensional fluid flow in a pipe.

Figure 11.1 represents a pipe with fluid flowing steadily through it. At plane 1 there is a velocity v, a density ρ, and an area A. The same variables exist at plane 2. All three terms may vary, depending on the substance and the flow conditions. The mass flow rate at any plane is

$$\dot{m} = \rho A v = \rho_1 A_1 v_1 = \rho_2 A_2 v_2 = \text{constant}. \qquad 11.1$$

The first law of thermodynamics is the postulate that energy is conserved; it can change form in a variety of ways, but its total value will be unaltered. Before we can write the most rudimentary equations describing this, we must define what is meant by energy. What do we call the device or substance undergoing, performing, or receiving the energy transformation?

11.3 SYSTEMS

There are two types of thermodynamic systems, open and closed. An open system allows mass to flow in and/or out. An automobile engine is an example of an open system; air and fuel flow into it, exhaust flows from it. The engine, an open system, requires a flow of matter through it. Figure 11.2 illustrates an automotive engine, with the various mass flows indicated. A closed thermodynamic system is a constant-mass system, illustrated by the heating of 2 kg of

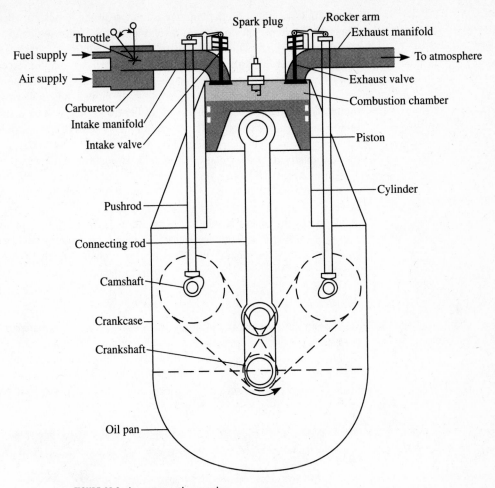

FIGURE 11.2 An automotive engine.

water. The 2 kg of water remain constant throughout the heat addition process. The reason for specifying these different systems is that the equations describing the conservation of energy are different for closed and open systems.

11.4 ENERGY FORMS

Matter possesses three energy forms, kinetic, potential, and internal energy. Two energy forms may enter or leave a system, work and heat. Each of these forms is described below.

Work

Work is a force F, acting through a distance Δx, or

$$W = F\Delta x,$$

11.2

where the force might be the horizontal force required to push a wheelbarrow a distance. The units of work are newton meters (N·m), or joules (J), in SI and foot-pound force (ft-lbf) in the English system. The force can take many forms. It can be the force acting on a mass to raise it, or it can be the force necessary to move a charged particle in a magnetic field. It may be a pressure acting on area, such as a piston crown, causing it to move. When work is used in thermodynamics, a system is involved. Either the system will be doing the work on the surroundings (everything external to the system) or the surroundings will be doing work on the system. To distinguish the two cases, we refer conventionally to work done by a system as positive and to work done on a system as negative. Notice that a system does not possess work; work results from an interaction between the system and the surroundings, for instance, a piston moving. The energy that allows the system to do work comes from energy contained by the matter within or passing through the system. Thus, work is an energy form that exists only in transition across the system boundary. Once it enters or leaves the system its effect causes a change in the matter's energy forms.

Heat

Heat is similar to work in that it is not an energy form that matter possesses; it is defined as energy crossing a system's boundary because of a temperature difference between the system and the surroundings. This definition differs from the colloquial usage of heat, in that matter does not have any energy form called heat. The sign convention for heat is opposite to that for work: heat flow into the

These parabolic tracking mirrors are used to focus the sun's energy at a central location, where the high-temperature energy is used as the heat source in a power plant. (Courtesy of ASME)

system is positive, heat flow from the system is negative. Heat is represented by the symbol Q. If there is a process in which $Q = 0$, the process is *adiabatic*. For instance, if a pipe is well insulated from the surroundings, then any heat flow between the system and the surroundings is very small, in fact, negligible in most cases. If it were zero, the pipe would be adiabatically insulated. The units of heat are joules or British thermal units (Btu). Notice a temperature difference still exists for the adiabatic case. Adiabatic conditions imply nothing about temperature.

Convertibility of Energy

We have seen that both foot–pound force and the British thermal unit are energy forms, but how are they related? Starting in 1845, an Englishman, James Prescott Joule, initiated experiments to find the equivalences between heat (Btu) and work (ft-lbf). Figure 11.3 diagrams a simplified version of his experiment. A closed, insulated container is fitted with a spindle to which paddles are attached. A wire is wound around the external spindle and led over a pulley to a weight (force). The weight is dropped at constant velocity, eliminating all acceleration except local gravity. The water temperature increases after the weight is dropped. From the definition of a British thermal unit at the time, the heat required to raise 1 lbm of water from 59.5° to 60.5°F, Joule was able to calculate the equivalence between mechanical energy and thermal energy. This has been refined until today, when 778.16 ft-lbf = 1 Btu. In SI both energy forms are joules, and the joule, British thermal unit, and foot–pound force are related by the following conversions:

$$1 \text{ Btu} = 1055 \text{ J} = 1.055 \text{ kJ}$$

$$1 \text{ ft-lbf} = 1.355 \text{ J}.$$

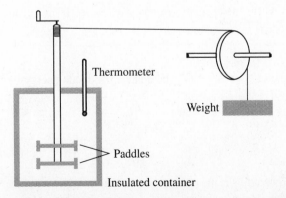

FIGURE 11.3 Joule's experiment on the equivalence of heat and work.

Potential, Kinetic, and Internal Energies

Matter may be modeled to contain many types of energy forms, but we will discuss only three: gravitational potential, kinetic, and internal energies—the ones that occur in the typical situations we will be analyzing.

The potential energy of a system mass depends on its position in the gravitational force field. There must be a reference datum and a distance z from the datum to the system mass. If a force F raises the system, the change in potential energy is equal to the work (force through a distance) necessary to move the system.

$$\Delta(\text{P.E.}) = F\Delta z = mg\Delta z$$

We assume that the value of g is constant, or that an average value is used over the Δz distance; thus,

$$(\text{P.E.})_2 - (\text{P.E.})_1 = mg(z_2 - z_1) \text{ N} \cdot \text{m} = \frac{mg}{g_c}(z_2 - z_1) \text{ ft–lbf.} \quad 11.3$$

We see that our assumption about the form of this energy equation is realistic: the potential energy of a system increases as the height increases.

The kinetic energy of a system is developed in an analogous manner. Again let us consider a system of mass m. A horizontal force acts on the system and moves it a distance x. Since it moves horizontally, there is no change in potential energy. The change in kinetic energy is defined as the work in moving the system a distance Δx.

$$\text{K.E.} = F\Delta x$$

$$F = ma = m \cdot \frac{\Delta v}{\Delta t}$$

where

$$\frac{\Delta v}{\Delta t} = \frac{\Delta x}{\Delta t} \cdot \frac{\Delta v}{\Delta x} = \bar{v} \cdot \frac{\Delta v}{\Delta x}$$

and where the average velocity $\bar{v}$ is defined from Equation 10.15 as

$$\bar{v} = \frac{v_1 + v_2}{2}.$$

Because $\Delta v = v_2 - v_1$, the change of kinetic energy becomes

$$\Delta(\text{K.E.}) = \frac{m}{2}(v_2 - v_1)(v_2 + v_1) = \frac{m}{2}(v_2{}^2 - v_1{}^2)\text{J}$$

$$= \frac{m}{2g_c}(v_2{}^2 - v_1{}^2)\text{ft-lbf.}$$

11.4

Note that this is the translational kinetic energy (au) of the system; if the velocity of the system increases, kinetic energy increases. If a car is moving at 40 km/h and is accelerated (force through a distance) to 80 km/h, the kinetic energy increases. If the car's velocity is zero, its kinetic energy is zero.

One of the less tangible forms of energy of a substance is its internal energy. This is the energy associated with the substance's molecular structure. Although we cannot measure internal energy, we can measure changes of internal energy. The symbol for internal energy is u, for total energy, U.

$$u = \text{specific internal energy, J/kg (Btu/lbm)} \qquad 11.5a$$

$$U = mu, \text{ total internal energy, J (Btu)} \qquad 11.5b$$

The change of internal energy for a substance may be written as

$$\Delta U = mc(T_2 - T_1), \qquad 11.6$$

where c is a property called specific heat. Typical values for specific heat are given in Table 11.1.

TABLE 11.1   SPECIFIC HEATS FOR VARIOUS SUBSTANCES

Substance	kJ/kg · K	Btu/lbm · R
Air	0.7176	0.1714
Aluminum	0.963	0.23
Brick	0.9210	0.22
Bronze	0.4353	0.104
Concrete	0.6530	0.156
Gasoline	2.093	0.50
Glass	0.8330	0.199
Ice	1.9883	0.475
Steel	0.419	0.10
Water (liquid)	4.186	1.00
Water (vapor)	1.4033	0.3352
Wood	2.51	0.60

11.5 THE FIRST LAW OF THERMODYNAMICS

The conservation of energy, or first law of thermodynamics for a constant mass system, may be developed as follows. Assume a system of constant mass receives heat from the surroundings and does an amount of work, and in the process its energy changes. Let $E = U + \text{K.E.} + \text{P.E.}$ represent the total energy that the substance has. Thus, postulating that energy is conserved may be stated by the word equation

$$\text{energy initial} + \text{energy in} = \text{energy final} + \text{energy out}$$
$$E_1 \quad + \quad Q \quad = \quad E_2 \quad + \quad W. \qquad 11.7$$

This may be combined to form

$$Q = E_2 - E_1 + W. \qquad 11.8$$

If we divide Equation 11.8 by the mass m, the quantities have units of energy per unit mass (kJ/kg, Btu/lbm), which is denoted by use of lowercase letters:

$$q = e_2 - e_1 + w.$$

If we expand e in terms of internal, kinetic, and potential energies, the following results for the first law, closed systems:

$$q = (u_2 - u_1) + \frac{v_2^2 - v_1^2}{2} + g(z_2 - z_1) + w. \qquad 11.9$$

■

E X A M P L E
11.1

An adiabatic tank contains 2 kg of water at 20°C and receives 20 kN · m of energy from a paddle wheel. Determine the final temperature.

SOLUTION:

We can use Figure 11.3 as a sketch of the problem. Determine from the given information whether or not terms may be eliminated from the first law equation. The tank is not moving; hence the kinetic energy and the change of kinetic energy are zero. The tank is not falling; hence the change of potential energy is zero. The tank is adiabatic; hence the heat loss is zero ($Q = 0$). Eliminating these terms from Equation 11.8,

$$0 = \Delta U + W$$

$$W = -20 \text{ kN} \cdot \text{m} = -20 \text{ kJ}$$

(work goes into the system and is negative).
    From Equation 11.6,

$$\Delta U = 20 = mc(T_2 - T_1) = (2)(4.186)(T_2 - 293)$$

$$T_2 = 295.4 \text{ K} = 22.4°C.$$

■

In using any of the first law equations, remember the sign convention for heat and work and substitute the energy value with the appropriate sign into the equation.
    The notation varies when energy per unit mass is multiplied by mass flow rate; and new energy term is per unit time, or

$$\dot{Q} = \dot{m}q \quad \text{W} \qquad (\text{Btu/min}),$$

and

$$\dot{W} = \dot{m}w \quad \text{W} \qquad (\text{Btu/min})$$

where $\dot{Q}$ is called the heat flux and $\dot{W}$ is called power.

The conservation of energy for an open system is given by the following expression for steady-state situations where heat flow is assumed into the system and work flow from the system:

$$\text{energy in} = \text{energy out} \qquad\qquad 11.10$$

$$\dot{Q} + \dot{m}_{in}(e + p/\rho)_{in} = \dot{W} + \dot{m}_{out}(e + p/\rho)_{out}.$$

For liquids the p/ρ term may be neglected, but not for gases.

E X A M P L E
11.2

A steam turbine, Figure 11.4, receives steam at a pressure of 900 psia and at a velocity of 100 ft/sec. The steam leaves the turbine at 1.5 psia and with a velocity of 900 ft/sec. The turbine entrance is 10 ft above the exit level. Find the work output if the mass flow rate is 100,000 lbm/h and the heat loss is 50,000 Btu/h. The steam has the following properties:

	Inlet	Exit
Pressure	900 psia	1.5 psia
Temperature	1000°F	114°F
Velocity	100 ft/sec	900 ft/sec
Specific internal energy	1354.5 Btu/lbm	951 Btu/lbm
Density	1.078 lbm/ft^3	0.0046 lbm/ft^3

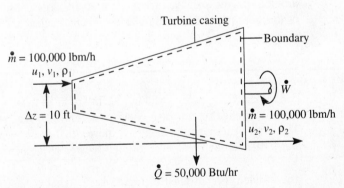

FIGURE 11.4 A turbine with steam flowing through it.

SOLUTION:

There is a mass flow into and out of the steam turbine; hence it is an open system. The first law for open systems, Equation 11.10, after dividing by the mass flow rate $\dot{m}$, becomes

$$q + u_1 + p_1/\rho_1 + (\text{K.E.})_1 + (\text{P.E.})_1 = w + u_2 + p_2/\rho_2 + (\text{K.E.})_2 + (\text{P.E.})_2.$$

Solving for w yields

$$w = (u_1 - u_2) + (p_1/\rho_1 - p_2/\rho_2) + [(\text{K.E.})_1 - (\text{K.E.})_2]$$
$$+ [(\text{P.E.})_1 - (\text{P.E.})_2] + q.$$

$$q = \dot{Q}/\dot{m} = \frac{-50,000}{100,000} = -0.5 \text{ Btu/lbm}$$

$$w = (1354.5 - 951) + \left[\frac{(900)(144)}{(1.078)(778)} - \frac{(1.5)(144)}{(0.0046)(778)} \right]$$

$$+ \left[\frac{100^2 - 900^2}{(2)(32.174)(778)} \right] + \frac{10}{778} - 0.5$$

$$w = 403.5 + 94.2 - 16 + 0.01 - 0.5 = +481.2 \text{ Btu/lbm}$$

$$\dot{W} = \dot{m}w = (1 \times 10^5)(481.2) = 4.812 \times 10^7 \text{ Btu/h} = 18,908 \text{ hp.}$$

■

11.6 FURTHER EXAMPLES OF ENERGY ANALYSIS

The utility of first law analysis is its application to a wide variety of energy consumption situations. The following examples illustrate some of these. By developing even rudimentary energy analysis techniques we better understand problems facing society, such as excessive energy consumption and the effects of this consumption, including acid rain and nuclear waste.

■
E X A M P L E
11.3

A hydroelectric power-generating facility is to be developed. A site is selected where a change of elevation of 45 m occurs, and the river flow rate averages 227 m³/s. Determine the maximum power that can be generated by having the water pass through hydraulic turbines located at the base of the dam.

SOLUTION:

The problem is illustrated in Figure 11.5. What is happening physically? The potential energy of the water at the higher elevation converts to kinetic energy prior to entering the turbine, and is then converted into work. The initial and final velocities of the water can be assumed to be negligible, and there is no heat transfer. The change of internal energy of the water is zero as well, because the energy transfer is between work and potential energy. This becomes apparent when we write the first law for this situation:

$$\text{energy in} = \text{energy out}$$

$$\dot{Q} + \dot{m}[u + p/\rho + (\text{K.E.}) + (\text{P.E.})]_{\text{in}} = \dot{W} + \dot{m}[u + p/\rho + (\text{K.E.}) + (\text{P.E.})]_{\text{out}}.$$

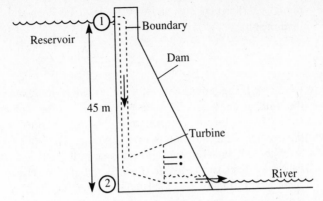

FIGURE 11.5 A hydraulic turbine used in the generation of electricity. The dammed water from a reservoir passes through the turbine, causing it to rotate and turn an electrical generator.

Water is incompressible, so the density is constant, nominally 1000 kg/m³, and the pressure in and out is the same, atmospheric. Therefore the p/ρ term adds out. Solving for the power yields

$$\dot{W} = \dot{m}[(\text{P.E.})_{in} - (\text{P.E.})_{out}].$$

The problem statement does not give us the mass flow rate, but rather the volume flow rate. Convert this to mass flow using the density.

$$\dot{m} = (1000 \text{ kg/m}^3)(227 \text{ m}^3/\text{s}) = 2.27 \text{ } 10^5 \text{ kg/s}$$

$$[(\text{P.E.})_{in} - (\text{P.E.})_{out}] = g(z_1 - z_2) = (9.8 \text{ m/s}^2)(45 \text{ m}) = 441 \text{ J/kg}$$

$$\dot{W} = (2.27 \times 10^5 \text{ kg/s})(441 \text{ J/kg}) = 100.1 \text{ MW}$$

The maximum power produced is 100.1 MW. Actually, this is reduced by several factors that may be combined into an overall efficiency term. Efficiency is a measure of output over input, of desired effect divided by the cost of the effect. In this case what is desired, the output, is the power produced. The cost of achieving this is the expenditure of the maximum power, the total energy available if all processes worked ideally. If the overall efficiency is 60 percent, then the actual power produced is

$$\eta = 0.60 = \dot{W}_{actual}/100.1$$

$$\dot{W}_{actual} = 60.1 \text{ MW}.$$

∎

EXAMPLE
11.4

A student wishes to investigate the space requirements for generating power using solar heat. She realizes that the solar radiation must be collected and concentrated using parabolic collectors. In addition she learns that the maximum possible efficiency for any cycle producing power can be given by the relationship

$$\eta_{Th} = 1 - \frac{T_C}{T_H} = \frac{\dot{W}_{net}}{\dot{Q}_{in}},$$

where T_H is the temperature of the high temperature source, the temperature of the fluid in the parabolic collector, and T_C is the ambient temperature of the surroundings, the air temperature in this case. Further investigation shows that the average solar radiation is 1.36 kW/m^2 = 430 Btu/h-ft^2 and that there are on average 3000 hours of sunshine per year in the area she lives. The municipality where she lives would like to produce part of its electrical consumption through such a solar plant. The total output power of 20 MW is desired. The fluid temperature in the parabolic collector is 200°C, and the ambient temperature averages 25°C. Determine the minimum land area required.

SOLUTION:

Figure 11.6 sketches the situation. We know the power output is 20,000 kW. From the cycle thermal efficiency we can determine the heat that is required.

$$\eta_{Th} = 1 - \frac{298}{473} = 0.37 = \frac{20,000}{\dot{Q}_{in}}$$

$$\dot{Q}_{in} = 54,054 \text{ kW}$$

The municipality wants the power delivered over 24 hours. While it is impossible to generate power during the night and on overcast days, the assumption is that excess power may be sold during the hours of operation and repurchased during the other hours. In this situation there are 3000 hours per year of operation.

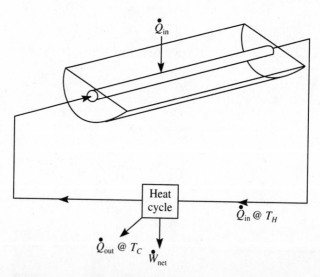

FIGURE 11.6 A parabolic solar collector focuses the sun's rays, creating a high temperature (T_H) heat source $\dot{Q}_{in}$. This energy may be used in a heat cycle to generate power, $\dot{W}_{net}$.

First let's find the annual kilowatt hours for the heat supply requirement.

$$\dot{Q} = (54,054 \text{ kW})(24)(365) = 4.735 \times 10^8 \text{ kwh/year}$$

Where does this come from? From the incident solar radiation with an intensity of 1.36 kW/m^2.

$$4.735 \times 10^8 \text{ kwh} = (1.36 \text{ kW/m}^2)(3000 \text{ h/year})(A \text{ m}^2)$$

$$A = 116\ 057 \text{ m}^2 = 28.67 \text{ acres}$$

∎

This is certainly a staggering land area to cover with collectors. An inherent problem with large-scale solar electric power generation schemes is the large amount of land required. This is not the case for home hot-water heating systems, where often partial coverage of the roof with collectors is sufficient area.

The following example analyzes the fuel requirements for a power plant that converts the chemical energy of the fuel source into thermal energy and thence into electrical power. The efficiency is given by the same equation as in Example 11.4. Fuels have different characteristics, one of the primary ones being the heating value of the fuel, the maximum heat that can be released during the combustion process of a unit mass of fuel. Table 11.2 lists some typical heating values for commonly used fuels.

TABLE 11.2 AVERAGE HEATING VALUES FOR VARIOUS FUELS

Fuel	Btu/lbm	KJ/kg
Coal	12 000	27 900
Corn cobs (dry)	9 300	21 600
Gasoline	19 260	44 800
Lignite	11 400	26 500
Natural gas	24 700	57 450
Residual oil	18 500	43 000
Wood (dry)	8 750	20 350

EXAMPLE
11.5

A power plant uses residual oil for fuel. The plant produces 1000 MW of electrical power, and the cycle producing the power has a high temperature of 550°C and a low of 25°C. Determine the daily fuel consumption.

SOLUTION:

Figure 11.7 shows a simplified power plant. The heat is supplied by burning fuel and generating steam at 550°C. The steam leaves the steam generator (boiler) and enters the turbine, producing work. Some of the work goes to the pump, but most

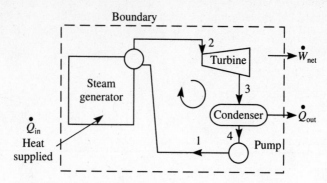

FIGURE 11.7 A simplified power plant.

leaves the system as the net work, power, produced turning the electrical generators that are attached to the turbine. The steam leaves the turbine and enters the condenser at 25°C and is condensed to a liquid and pumped back to the steam generator. The heat removed in the condenser, $\dot{Q}_{out}$, is the difference between the heat supplied, $\dot{Q}_{in}$, and the power produced, $\dot{W}_{net}$. Thus, $\dot{Q}_{out} = \dot{Q}_{in} - \dot{W}_{net}$, where the heat out is written as a positive value, ignoring the thermodynamic sign convention.

$$\eta_{Th} = 1 - \frac{298}{823} = 0.638 = \frac{1000 \text{ MW}}{\dot{Q}_{in}}$$

$$\dot{Q}_{in} = 1567 \text{ MW}$$

$$\dot{Q}_{in} = 1\,567\,000 \text{ kW} = (\dot{m}_f \text{ kg/s})(43\,000 \text{ kJ/kg})$$

$$\dot{m}_f = 36.44 \text{ kg/s}$$

$$\dot{m}_f = (36.44 \text{ kg/s})(24 \text{ h/day})(3600 \text{ s/h}) = 3\,148\,416 \text{ kg/day}$$

Determine the volume of a cylindrical storage tank and its dimensions if $L = D$ and it must hold 3 days' supply of fuel. The density of residual fuel is 990 kg/m.

$$V_f = (36.44 \text{ kg/s})(24 \text{ h/day})(3600 \text{ s/h})(990 \text{ kg/m}^3)(3 \text{ days})$$

$$V_f = 9540.6 \text{ m}^3 = \frac{\pi}{4}D^2L = \frac{\pi}{4}D^3$$

$$D = 22.99 \text{ m}.$$

◼

Very often residual fuel and coal have constituents that when burned cause us problems, like sulfur. In the combustion process the sulfur forms sulfur dioxide, which combines with water, such as rainwater, to form sulfuric acid. This is one of the main ingredients of acid rain. The carbon dioxide formed in the combustion process hastens the greenhouse effect, or warming of the Earth. We

This steam turbine, when assembled and operating, will be able to generate tens of thousands of horsepower. The steam moves through a set of blades, attached to the rotor, then through nozzles located in the turbine casing and back again to another set of blades. The force of steam moving across the turbine blades causes the turbine to rotate and in turn rotate an electric generator attached to it. (Courtesy of Trident Engineering Associates)

The white plume emitted in cool weather from smokestacks is not smoke, but condensed water vapor. In the combustion of hydrocarbon fuels, the hydrogen combines with oxygen to form water. When the water vapor leaves the stack and is cooled sufficiently by the surrounding air, it turns to a fine mist. (Courtesy of Trident Engineering Associates)

can calculate the magnitude of these terms quite easily. The gravimetric analysis of the fuel tells us the percentage of the various constituents in each unit mass of fuel. For instance, fuel oil may contain carbon, hydrogen, sulfur, water, and ash. Let's say that the fuel contains 1 percent sulfur and that the cleaning processes on the stack remove 98 percent of the sulfur dioxide from the combustion gases. Calculate the kilograms of sulfur dioxide that are produced from this one power plant per day.

Initially let's determine the total mass of sulfur that will be emitted from the stack of the power plant in Example 11.5. The daily mass of fuel burned is

$$\dot{m}_f = 3\ 148\ 416 \text{ kg/day};$$
$$\dot{m}_S = (0.01)(3\ 148\ 416) = 31\ 484 \text{ kg/day}.$$

However, 98 percent of this is contained in the stack, so only 2% of this value leaves the stack

$$(\dot{m}_S)_{emitted} = 630 \text{ kg/day}.$$

The sulfur reacts with oxygen to produce sulfur dioxide according to the following reaction equation:

$$S_2 + O_2 = SO_2$$
$$32 \text{ kg S} + 32 \text{ kg O}_2 = 64 \text{ kg SO}_2.$$

The reaction equation is simply a conservation of mass, as indicated above. In this case one mole of sulfur, 32 kg, reacts with one mole of oxygen, 32 kg, to form one mole of sulfur dioxide, 64 kg. Thus for every 32 kg of sulfur burned, 64 kg of sulfur dioxide are formed. Let's just consider the sulfur that escapes.

$$\frac{64 \text{ kg SO}_2}{32 \text{ kg S}} \times 630 \text{ kg/day S} = 1260 \text{ kg/day SO}_2 \text{ released}$$

This amounts to 1.39 tons per day from this one plant. Imagine the amount that would be released if scrubbers were not installed. Not all industries, nor all countries, use exhaust gas scrubbers.

11.7 THE SECOND LAW OF THERMODYNAMICS

The thrust of the thermodynamic analysis in this text will be confined to the first law. However, as you progress through your engineering career, aspects of the second law will become very important. The second law of thermodynamics is the only physical law that is not an equality. It establishes an energy quality level, a value system for energy. Essentially, the more useful energy is in its ability to help us, the greater its value. Thus, energy at a high temperature is more valuable than energy at a low temperature. For instance 1000 kJ of energy at 35°C will warm us, but 1000 kJ of energy in the ocean at 4°C will not. It is less valuable to us. Additionally, the second law determines the direction of energy flow, from a higher energy quality to a lesser energy quality. Thus, energy flows from a high temperature to a low temperature. The law establishes the maximum efficiency for power production and can indicate whether a chemical reaction can occur in a given situation.

The second law may be stated as follows:

> Whenever energy is transferred, the level of energy cannot be conserved, and some energy must be permanently reduced to a lower level.

This may be combined with the first law:

> Whenever energy is transferred, energy must be conserved, but the level of energy cannot be conserved, and some energy must be permanently reduced to a lower level.

Availability, the energy available to do work for us, is the thermodynamic quantity that is most frequently used to denote energy quality. Consider the following situation. Fuel is burned in a home heating furnace. The products of combustion are at a high temperature, and the energy associated with it has a high availability. Energy is transferred from the products of combustion via heat transfer to water for use in the house radiators. The water's availability is less than that of the products of combustion. The radiators transfer energy to the room, raising the room's air temperature. The air's availability increases. In each transfer process, combustion gas to water, water to air, there was a net decrease in the total availability. The availability of the products of combustion decreased much more than the gain of availability by the water. Similarly, availability decreases in the heat exchange between the water and air. Finally, energy is exchanged between the house and the surroundings. The availability of energy at the temperature of the surroundings is zero, it can do no work for us. Thus, the total energy released by the combustion process, let's say 1000 kJ of heat which have an availability of 700 kJ, eventually are transferred to the surrounding environment. The first law states that 1000 kJ must reach the surroundings, and it does. The second law tells us that at that time the availability will be zero.

It is important to conserve energy and energy quality. Conserving energy is made possible by using less for a given task or improving the first law efficiency for a process—for example, turning off extra lights or driving cars with higher gas mileage. Additionally, it is possible to conserve energy quality by matching source and system energy quality needs, improving the second law efficiency of the process. The second law efficiency relates the availability requirements of the system using the energy to the availability attributes of the source providing the energy. The higher the efficiency, the better the match. Using energy of high quality, namely, availability, in those situations requiring it assures that people in the near future will have the necessary energy resources. We should examine practices such as burning oil, which has a high availability, for a situation like space heating, which requires heat with a low availability.

11.8 ENERGY AND OUR FUTURE

We love energy. Our lifestyle and our standard of living depend on a very high rate of consumption of energy. We forget that all the electricity and gasoline come from a finite supply of fossil fuels. Because it is a finite supply, it will be depleted, depleted very soon, within our lifetimes for domestic oil and gas reserves. Figure 11.8 illustrates the energy sources within the continental United States as of 1980. This does not indicate where the fuel reserves are; the United States has a large coal supply, but the domestic petroleum and natural gas supplies are decreasing, requiring that we import more of them. Various factors are weighed to estimate when the domestic supplies will be depleted and when we will become totally dependent on foreign supplies. Certainly in the next 10 to 25 years the United States will be in this situation. Economic and political problems develop as a result of this dependency: economic in that the cost of fuel will rise dramatically, causing inflation and social pressure, and political in that other countries will be vying for the same energy sources. Although this section is directed toward the technical implications of changing energy sources, as engineers we must be very aware of the social and political pressures associated with changes as well. Figure 11.9 represents a forecast of the changing proportions of the energy sources in the United States for the next 40 years. It is apparent that oil and gas will decrease significantly and that coal will be the remaining fossil fuel, with nuclear, solar, and hydroelectric power being the other energy sources.

The United States has large coal reserves that can be used for power generation, coal gasification projects, and coal liquefaction processes. Studies have indicated that there is sufficient coal to provide for our energy needs for the

Much of the U.S. electrical energy is generated at plants such as this one. Engineers design not only the generating facility, but the storage facilities and transportation system for bringing the fuel to the power plant. (Courtesy of Public Service Electric and Gas)

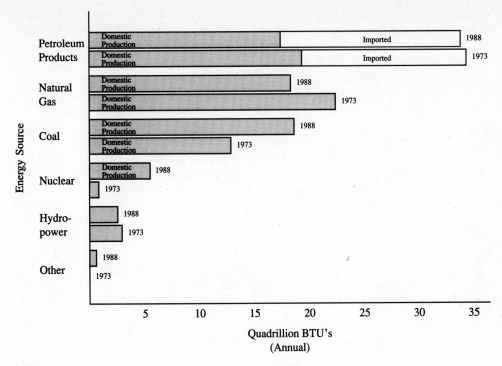

FIGURE 11.8 The energy sources for the United States in 1980.

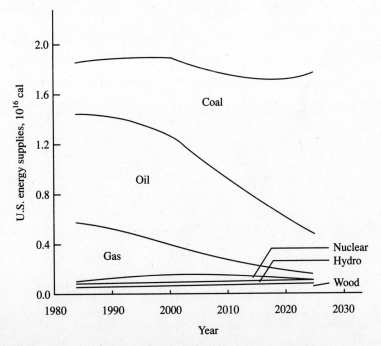

FIGURE 11.9 A projection of changes in U.S. energy sources with time.

next 400 years. However, there are tremendous environmental problems associated with expanding our use of coal. Much of the coal is accessible by strip mining, a technology that arouses environmental group action and requires water, water that is needed for agriculture and drinking in the same geographic regions. Of the coal reserves 20 percent are high-sulfur coals nationwide, with 43 percent high-sulfur coal in the coal regions east of the Mississippi River. The high sulfur content means the coal must be processed to remove as much sulfur as possible. (Remember the correlation between sulfur and acid rain.)

Nuclear power generation is at a standstill in the United States. No more plants are being constructed, as the legal impediments to building a plant are virtually insurmountable. Nuclear plants create significant waste disposal problems, one of the key areas for concern. Internationally, however, nuclear plants are being constructed as an alternative power source, and we may see them appear in the United States once again if the time for construction can be decreased.

Renewable energy supplies, solar and hydro, hold some promise for future development, particularly small-scale solar power. Hydroelectric power has severe environmental impacts associated with it. Dams created for it flood land,

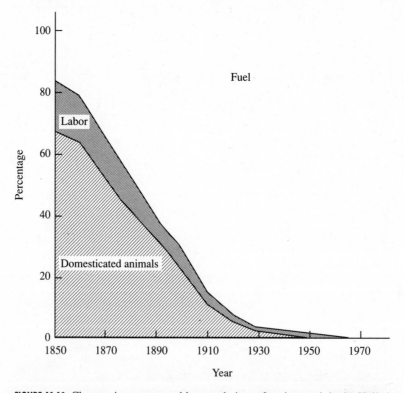

FIGURE 11.10 Changes in energy used by people in performing work in the United States. From C. A. S. Hall, C. J. Cleveland, and R. Kaufmann, *Energy and Resource Quality: The Ecology of the Industrial Process* (New York: John Wiley and Sons, 1986).

change water tables, and destroy the habitat of certain species. Dams can still be constructed, but it takes time and is sure to engender significant opposition. Certain types of solar energy utilization, particularly passive solar energy for home heating and photovoltaics for electric power generation, will increase with time.

How did we get in this predicament in the first place? As we have progressed through the industrial age, fuel has everywhere replaced people and animals in performing work in manufacturing, agriculture, transportation. Figure 11.10 graphs the change in energy source for work over time. Energy has lifted the burden of toil from our backs. People use machines (energy provided by fuel) for manufacturing jobs, be it a riveter or a computer assembler. Figure 11.11 shows the increase of productivity that is associated with this increased energy utilization. As we reach a terminus in the nonrenewable fuel supplies, we must rethink productivity gains and search for them by means other than energy. Engineering creativity is certainly needed. Figure 11.12 illustrates the near 100 percent dependence on nonrenewable energy sources in the United States. Even using foreign supplies to replace domestic sources, everything else being the same, we must change our consumption habits, as all fossil fuel supplies are finite.

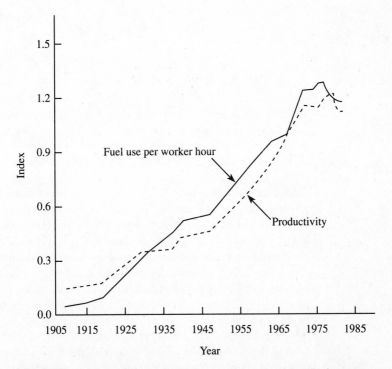

FIGURE 11.11 Changes in energy usage and worker productivity. Both measures are indexes, 1967 = 1.0. From C. J. Cleveland, R. Costanza, C. A. S. Hall, and R. Kaufmann, "Energy and the U.S. Economy: A Biophysical Perspective," *Science* (1984) 225:890-897. © 1984 AAAS.

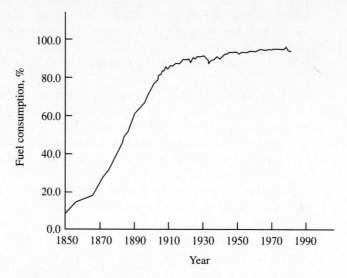

FIGURE 11.12 The percentage of fuel consumed in the United States that is nonrenewable.

Improving Energy Usage

Immediate action can be taken to improve energy utilization and conservation, both first and second law energy conservation. Because fuel costs have been very inexpensive, artificially so for a variety of reasons, the manufacture of products has tended toward lowest initial cost, not lowest total cost, with operating costs included. For instance, it is possible using current technology to build domestic refrigerators that use a maximum of 700 kwh per year rather than the maximum of 1900 kwh per year that current models use. California recently passed a law requiring that by 1993 all refrigerators achieve the greater efficiency. Engineers created new models with design changes requiring more insulation and more efficient compressors, motors, condensers, and evaporators. These new models cost only $50 more per unit. Super-efficient models are being constructed that annually consume 100 kwh of electricity, 5 percent of the 1900 kwh units.

Switching to natural and fluorescent lights from the predominant incandescent light bulb will save equally large amounts of electricity. Consider the following: a 75-watt incandescent light bulb lasts an average of 750 hours and produces 1180 lumens of illumination; a 20-watt fluorescent light lasts 7500 hours and produces 1250 lumens of illumination. Much, 90 percent or more, of the incandescent bulb's energy is dissipated as heat. The bulb functions by heating the filament to a high temperature; the filament glows, producing light. Of course, this heat must be removed by a building's ventilation system in hot weather. Fluorescent bulbs are coated with phosphor powders that glow when excited by ultraviolet light. The ultraviolet light is created by ionizing a gas within the tube.

Wise use of our energy resources is demonstrated by this farmer's use of wind power. Wind power is ideal for remote locations and for uses that can tolerate fluctuations or intermittent operation. Often a battery back-up system provides electrical energy in times of low wind speed. (Courtesy of ASME)

This industrial gas turbine can be used for a variety of purposes, such as driving electrical generators in times of peak demand. Electrical utilities often have gas turbine peak loading plants. The gas turbine can reach full load conditions in a matter of seconds and can be started from a remote location. (Courtesy of ASME)

In the early 1970s automobiles had a comparatively low fleet mileage average, 15–16 miles per gallon (mpg). Since the first energy crisis in 1973 the fleet average has risen to 25–26 mpg. This represents a tremendous savings in fuel when you consider the number of cars on the road and miles driven. However, some automobiles currently obtain 50 mpg or better. By decreasing vehicle size and weight and using new materials, more efficient engines, and better aerodynamic design, the fleet average can continue to rise.

Not only will our personal transportation systems be changed with decreased fuel availability and/or significantly higher fuel costs, but other systems will change as well. There has been a switch over the past 40 years from shipping goods by railroad to shipping by trucks. The days are numbered for this, for long-distance hauling trains are at least five times more fuel-efficient than trucks. Compare a freight train with one engine to the number of trucks that would be needed to haul the same goods.

Energy-Intensive Agriculture

Populations and societies prosper because of abundant food and energy supplies. Part of the reason we have been able to support our standard of living at its present level is that we export agricultural products, creating a trade surplus to partially counter the trade deficit in energy. Problems are lurking in this vital sector of the economy. Farming has developed into a highly energy-intensive operation, and farms have become larger as a result of this. Agricultural engineers will be needed to create new technologies and methodologies in the future to support this vital economic area. Certainly tractors are needed for tilling the soil, but this is not the problem, although techniques in plowing dictated by large tractors cause soil erosion, the loss of a nonrenewable resource. Synthetic fertilizers, rather than naturally occurring nutrients, are used to increase crop yield per acre; these fertilizers are produced using natural gas. Pesticides and herbicides, also created using fuel, are applied to control insects and weeds. Of course, energy is expended in the repeated applications. The pesticides, herbicides, and fertilizers enter the water supply, causing pollution, as much farmland is irrigated, with the aforementioned chemicals becoming part of the runoff. The pollution of water supplies, rivers, lakes, even the ocean, is caused in part by this intensive energy application in growing our food supply.

Aggravating the situation is the finding that only 25 percent of the crops are for human consumption: the rest are for animals. It is currently cheaper to raise grain-fed animals rather than grazed animals. The meat is fatter and more tender and composes 35 percent of the typical American diet. As you can see, the intertwinings of various activities in support of our lives is quite complex, and the solutions are not easy or comfortable. However, they are feasible.

Agricultural engineers are interested in the optimum lighting requirements for plant growth. Not only do plants respond to different wavelengths of light, they also respond to different periods of light and dark in achieving maximum growth. (Courtesy of Public Service Electric and Gas Company)

Energy Policy

Figure 11.13 graphs the energy consumption in various countries and the GNP per person in those countries. At one extreme is the United States and at the other is India. Energy consumption per person in the United States is about 80 kcal and in India is about 2 kcal. The standard of living, however, indicated by the GNP, is also significantly different. Let's look instead at countries whose standard of living is not quite as high as the United States, using the GNP as the only correlator, but whose energy consumption is less than half that of the United States. Countries such as Norway, France, Switzerland, and Germany offer energy-efficient methodologies that we may wish to use in the future to decrease our energy consumption habit.

Engineers are needed to create new technologies and modify existing ones so we may continue to prosper as a country. There are two fundamental energy philosophies that will be in the news, hard and soft energy policies. The hard energy policy favors large centralized power-generating facilities. The expansion

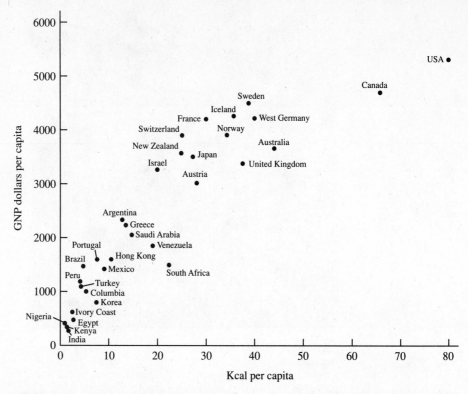

FIGURE 11.13 A comparison of the total energy use per capita and the gross national product (GNP) dollars per capita for various countries. From C. A. S. Hall, C. J. Cleveland, R. Kaufmann, *Energy and Resource Quality: The Ecology of the Industrial Process* (New York: John Wiley and Sons, 1986).

of nuclear power, creation of large hydroelectric power facilities, and development of coal gasification facilities typify this policy. The soft energy policy advocates small solar power systems, cogeneration facilities, and increased use of renewable energy supplies. Both policies have problems associated with them, although the soft energy policy has an innate long-term appeal, considering environmental factors.

Problems 11.32–11.36 illustrate the enormous effect that small changes can have on the national energy consumption. A few of the many examples are use of more efficient motors in appliances, insulation of homes and buildings, and reduction in the speed limit; all result in dramatic fuel savings.

Photovoltaics

One of the most promising renewable energy forms that is sure to become a significant electrical energy supplier in the future is photovoltaics, the direct conversion of sunlight into electricity. Current laboratory models are able to attain a conversion efficiency of 18 percent on small cells and an average

Technological advances in battery storage systems have not kept pace with advances in other areas of electrical engineering. This engineer is developing and testing new zinc chloride batteries. (Courtesy of Public Service Electric and Gas)

efficiency of 13 percent over a square foot of area. Production costs are also dropping, making the units more economically attractive.

How does this conversion occur? Photovoltaic cells are a type of semiconductor involving a P-N junction. A P-N junction has a voltage potential created by a positive-hole P-layer, which contains movable positive charges or "holes," and a negative N-layer, which contains movable negative electrons. When light with sufficient energy enters the crystal, electrons are released from their atomic bonds and migrate to an electrode. A wire connected from the electrode to a load leads back to the positive electrode, where the electrons combine with the positive holes. A barrier at the P-N junction prevents the electrons from instantly combining with the holes in the P-layer, forcing them instead to flow from the electrode. No material is consumed, and the cell can theoretically last forever, or until other parts wear out.

Let's use a silicon cell to illustrate the current flow. Pure silicon has a valence of four, and the atoms are arranged in a regular and uniform lattice. If a five-valence atom, such as arsenic, is introduced into the lattice structure replacing one of the silicon atoms in the lattice, four of its electrons will bond the silicon atoms, leaving the fifth electron free to move and be a carrier of electricity. This is used to create the N-layer. To create the P-layer, a three-valence atom is introduced into the silicon lattice. However, it satisfies only three of the four bonding requirements of the silicon atom, leaving a positive hole in the crystalline structure. (Adding impurities to a pure crystalline structure is called "doping.")

N-layer with
fixed positive
holes in a silicon
lattice. Electrons
are free to move.

Barrier layer
with fixed holes
and electrons.

P-layer with
fixed electrons in
a silicon lattice.
Positive holes are
free to move.

FIGURE 11.14 A P-N photovoltaic cell.

Figure 11.14 illustrates a P-N photovoltaic cell. The barrier layer prevents diffusion of holes or electrons from the P-layer to the N-layer, or vice versa. As sunlight enters the crystal, it is absorbed, producing an electron and a hole. Ordinarily these would immediately recombine, and the effect of the light absorption would be an increase in crystalline temperature. However, because of the potential barrier at the P-N junction, the electrons migrate to one electrode and the holes to the other. This produces a potential difference, and a current can flow between the electrodes when a wire connects them. Very important to the successful operation of the cell is controlled doping with selected impurities and the absence of other impurities. The effect of other impurities is to allow recombination of holes and electrons, rather than a flow of current. Because of engineering advances in refining and doping, photovoltaic power generation is now a practical means of small-scale power generation.

Cogeneration

Another type of energy system you will be involved with more and more is a cogeneration system. Cogeneration means using the same energy source for more than one purpose, such as using the waste heat from an engine for space heating. Cogeneration facilities may be used to generate electricity locally. For example, a university might use its own diesel-electric generators. Then the waste heat from the engine is used for additional purposes, perhaps for space heating or for air conditioning, as odd as that sounds. Large air conditioning systems that are absorption refrigeration systems use heat as the energy source. So, rather than having an engine that is 40 percent efficient in generating electricity, with the waste heat from the cooling water and the exhaust being dissipated to the atmosphere, a cogeneration facility uses this waste heat for other purposes. This increases the overall efficiency of the unit and decreases the fuel or utility costs that the university has to pay.

You will be a practicing engineer in a challenging time; the energy supplies that we have become accustomed to will decrease dramatically. New technologies must be created to help us continue the advantages that society now offers us. Certainly our energy habits will change, natural and fluorescent lighting will become dominant, electricity generation will change increasingly to photovoltaic and cogeneration facilities, transportation patterns will alter, food production techniques will become less energy-intensive. All of these changes can be accomplished with talents in which engineers abound: creativity, compassion, and problem-solving ability.

REFERENCES

1. Burghardt, M. D. *Engineering Thermodynamics with Applications*. 3rd ed. Harper & Row, New York, 1986.
2. Daniels, F. *Direct Use of the Sun's Energy*. Ballantine Books, New York, 1964.
3. Gever, J., et al. *Beyond Oil*. Ballinger Publishing Co., Cambridge, Mass., 1986.
4. *State of the World 1987–1991, Worldwatch Institute Report*. W.W. Norton Company, New York, 1991.

PROBLEMS

11.1 A garden hose is 1 in. in diameter and has water with a density of 62.4 lbm/ft^3 flowing steadily through it with a velocity of 75 ft/sec. Determine the mass flow rate of water through the hose. Convert this to gallons per minute.

11.2 A steam turbine has an inlet steam flow of 4 kg/s with a density of 20 kg/m^3. The inlet diameter is 10 cm, and the outlet diameter is 20 cm. The outlet density is 10 kg/m^3. Determine the inlet and outlet velocities.

11.3 Water with a density of 1000 kg/m^3 flows steadily through a pipe with an internal diameter of 5 cm. The volume flow rate is 0.5 m^3/s. Determine the mass flow rate and velocity.

11.4 When an incompressible fluid flows through a pipe, the density by definition is constant. Show that for a given mass flow rate the velocity is inversely proportional to the square of the pipe diameter.

11.5 Two gaseous streams containing the same fluid enter a mixing chamber and leave as a single stream. For the first gas the entrance conditions are $A_1 = 500$ cm^2, $v_1 = 130$ m/s, $\rho_1 = 1.60$ kg/m^3. For the second gas the entrance conditions are $A_2 = 400$ cm^2, $\dot{m}_2 = 8.84$ kg/s, $\rho_2 = 1.992$ kg/m^3. The exit stream condition is $v_3 = 130$ m/s and $\rho_3 = 2.288$ kg/m^3. Determine the total mass flow leaving the chamber and the velocity of the second gas entering the chamber.

11.6   One hundred kJ/kg of heat is added to 10 kg of a fluid while 25 kJ/kg of work is extracted. Determine the change of internal energy. Find the temperature change if the substance is
a) water;
b) air.

11.7   A container holds 15 l of water. If 2000 kJ of heat is added, what is the temperature change?

11.8   An experiment is run to determine the specific heat of a certain material. The test shows that a 4-lbm block inreases 10°F when 8.4 Btu of heat is added. What is the material's specific heat?

11.9   An adiabatic pump casing holds 5 kg of water and delivers 2 kW of power to the water as it moves through the pump. The discharge valve of the pump is closed. The pump delivers the power to the water in the casing for 1 min. What is the water's temperature rise?

11.10 Two kg of boiling water (100°C) is poured into a 0.7-kg steel container at 20°C. What will their final equilibrium temperature be, assuming no losses to the surroundings?

11.11 Determine the energy release from burning
a) 1 ton of coal;
b) 1 gallon of gasoline (specific gravity = 0.835);
c) 5 kg of natural gas.

11.12 The burning of 0.75 kg of natural gas could propel a 1200-kg automobile at what velocity?

11.13 A tank with a diameter of 2 m and a height of 3 m is located 50 m above the ground. It is filled with water from a pump located on the ground. What energy is required? If the tank fills in 1 hour, what average power is required?

11.14 A 0.5-kg container is dropped from the top of a 75-m tall building. Determine its kinetic energy and potential energy at
a) the moment it is dropped;
b) after it has fallen 50 m;
c) the instant before hitting the ground.

11.15 An airplane weighing 10,000 kg is flying 2,000 m above the earth's surface at 1,000 km/h. Determine the plane's kinetic and potential energies.

11.16 A rifle bullet has a mass of 1.5 g and leaves the barrel of the gun at 500 m/s. Determine its kinetic energy.

11.17 Five people are lifted on an elevator a distance of 100 m. The work is found to be 343 kJ. Determine the average mass per person.

11.18 A student is watching pilings being driven into the ground. From the size of the pile driver, the student calculates its mass to be 500 kg. The distance that the pile driver is raised is measured to be 3 m. Determine the potential energy of the pile driver at its greatest height (the piling is considered the datum). Find the driver velocity just prior to impact with the piling.

11.19 A child (25 kg) is swinging on a play set. The swing's rope is 2.3 m long. Determine the change in potential energy from 0° with the vertical when the swing goes from 0° to 45° with the vertical, and the velocity when the swing again reaches the 0° angle.

11.20 An adiabatically insulated 2-kg container is dropped from a balloon 3.5 km above the earth. Upon impact with the ground the box remains intact; the volume remains the same, so no work is done on it. What is the change of internal energy of the box after impact?

11.21 A fluid enters a device with a steady flow of 3.7 kg/s, an initial pressure of 690 kPa, an initial density of 3.2 kg/m^3, an initial velocity of 60 m/s, and an initial specific internal energy of 2000 kJ/kg. It leaves at 172 kPa, ρ = 0.64 kg/m^3, v = 160 m/s, and u = 1950 kJ/kg. The heat loss is found to be 18.6 kJ/kg. Find the power (work per unit time) in kilowatts.

11.22 Calculate the kinetic energy of a 3000-lbm automobile moving at 60 mph.

11.23 The automobile in Problem 11.22 is stopped. The brakes have an average specific heat of 0.22 Btu/lbm-R. Assume that one-half of the energy is adiabatically absorbed by the brakes, which have a collective mass of 15 lbm. Determine the temperature rise of the brakes.

11.24 Steam with a flow rate of 3000 lbm/h enters an adiabatic nozzle at 200 psia, 600 ft/min, a density of 0.424 lbm/ft^3, and a specific internal energy of 1122.7 Btu/lbm. The exit conditions are p = 20 psia, ρ = 0.057 lbm/ft^3, and u = 973 Btu/lbm. Determine the exit velocity. (Note: no work is done in a nozzle: thermal energy is converted into kinetic energy.)

11.25 Refer to Example 11.3. It is now summer and the reservoir level has decreased to 35 m and the flow has been reduced to 135 m^3/s. Determine the power produced.

11.26 Refer to Example 11.4. The high temperature output of the parabolic collector increases to 210°C for 1500 of the 3000 hours of sunlight. Determine the power produced.

11.27 The south-facing roof of a house is 400 ft^2. Determine the maximum energy that can be captured if the annual hours of sunlight are 2900 and the radiation intensity is 400 Btu/h-ft^2.

11.28 The house in Problem 11.27 requires 1.295×10^8 Btu annually for home heating. The collector is 60 percent efficient. What is the minimum collector surface area required?

11.29 Refer to Example 11.5. Determine the heat leaving the power plant. If water is used for cooling, receiving the heat from the power plant, and it increases from 15°C to 25°C, what flow rate in gallons per minute is required?

11.30 The cooling water supply in Problem 11.29 comes from a lake where the return is mixed. Atmospheric cooling at night maintains a stable temperature. However, the specifications require that the lake must be large enough such that the mixing of the 25°C into the lake water at 15°C will not cause the lake water to increase in temperature more than 0.5°C in a 24-hour period. How large must the lake be? (Determine the mass required and then the volume, knowing the density of water is 1000 kg/m³.)

11.31 Refer to the power plant in Example 11.5. The fuel used is now coal with 3 percent sulfur. Determine the sulfur dioxide produced and the tons of coal required per day. If a railroad car holds 86,000 kg, how many carloads of coal are needed per week?

11.32 Compare the operating costs of a home refrigerator that uses 700 kwh electricity annually to one that uses 1900 kwh. The cost of electricity is $0.10 kwh. If this were enacted nationally, such that 10 million refrigerators were effected, what would be the total savings in kilowatt-hours?

11.33 Consider the fuel savings in Problem 11.32, because less electricity needs to be generated. A power plant may be 40 percent efficient in converting the fuel's chemical energy into electrical energy. Assume the fuel is oil with a heating value of 43,000 kJ/kg. How many kilograms would be saved annually?

11.34 Refer to the energy and lifetime comparisons between incandescent and fluorescent lights. Using a 75-watt incandescent bulb and a 20-watt fluorescent bulb, investigate the initial cost of each at a local store. Assuming that the light is used 8 hours per day annually, determine the time necessary to recover the additional cost of the fluorescent bulb.

11.35 Refer to Problem 11.34. Calculate the total energy savings over the lifetime of the fluorescent bulb. If nationally 10 million bulbs have been changed to fluorescent, calculate the fuel savings annually, using the same assumption as in Problem 11.33.

11.36 Consider a subset of the American automotive fleet that comprises one million cars that are driven 10,000 miles annually. The average gasoline consumption for this fleet rises from 26 mpg to 31 mpg over a 5-year period. Calculate the total fuel savings annually and cumulatively over this time.

11.37 The following data represent the proven world oil reserves as of 1989 in billions of barrels of oil:

Saudi Arabia	255	Venezuela	58.5
Iraq	100	USSR	58.4
United Arab Emirates	98.1	Mexico	56.4
Kuwait	94.5	United States	25.9
Iran	92.9	China	24.9

The following represents the consumption in millions of barrels of oil in 1989 of the largest consumers:

United States	6,323	France	677
Japan	1,818	Canada	643
West Germany	831	Great Britain	634
Italy	708		

Determine the number of years of oil supplies remaining, assuming consumption remains constant. Assuming the consumption increases at 3 percent per year, determine the number of years remaining.

12

ENGINEERING ECONOMICS

■ To define capital and learn about using profits.

■ To calculate the time value of money.

■ To evaluate the economics of alternative proposals.

■ To address the feasibility of public works projects.

(Photo courtesy of H. Armstrong Roberts.)

*E*ngineers are aware of production lines and bottom lines. Fundamental to a successful design is that it be economical to manufacture. Determining the value of money, now and in the future, is a skill engineers must hone in evaluating alternative designs.

12.1 INTRODUCTION

One of the most valuable courses you may take is one in engineering economics; it will be an asset to you in your engineering career and in your nonengineering life.

In chapter 1 we discussed the rudimentary beginnings of an engineering firm that builds bridges, the effect of competition, and the need to control costs. At that time we glossed over the economic analysis of the problem, as the thrust of the discussion was toward the various components of an engineering education and developing the identity of who an engineer is. Now that we have a better understanding of what is required in engineering decisions, we recognize that economic analysis is pivotal in determining design criteria, project cost, and product competitiveness.

Very important to companies is the economic comparison of design alternatives or leave equipment that is to be purchased. Suppose you are working for a small manufacturing company that needs to purchase a new machine to perform a variety of machining tasks. You have narrowed the search to two machines that can perform all the required tasks and are of good quality, produced by reputable manufacturers. However, the costs are different for set-up time between operations, operational requirements, expected lifetime of the machine, initial price, end-of-life value, maintenance, and power requirements. All of these factors must be modeled economically so that you can compare the two machines quantitatively.

12.2 CAPITAL AND PROFIT

Capital is another name for wealth, money, or property. Property has a certain conversion factor into money, so pragmatically capital and money are the same, though capital is philosophically more far-reaching. All businesses, engineering and otherwise, require capital for their operation. Expenses incurred must be paid before a project or product returns income to the business. Capital can be of two forms, equity capital, for example, money in a savings account, and borrowed capital, funds borrowed from another source, such as a bank. Owners of the borrowed capital receive interest from the company for the use of the money. Why borrow this money in the first place? Most often a company will not have enough money available to fund the costs of a new product or project; it has some and will borrow the rest. The company hopes it will receive income from the product or project, which will more than cover the costs, yielding a profit:

$$\text{profit} = \text{income} - \text{costs}. \tag{12.1}$$

Costs that are deducted from the income fall into the following general categories:

1. Costs associated with operation and maintenance ($O + M$) such as salaries, materials consumed, and utilities;
2. Interest paid on borrowed capital (I);
3. Depreciation of physical assets, such as the building, machines, or computers, used in the support of the product or project (D);
4. Taxes, which are based on the profit before taxes (T).

Let G be the gross income the company receives, PBT the profit before taxes, and PAT the profit after taxes.

$$\text{PBT} = G - (O + M + D + I) \tag{12.2}$$

and

$$\text{PAT} = \text{PBT} - T. \tag{12.3}$$

We are familiar with efficiencies (output over input), and a common financial efficiency is rate of return, which is the annual net profit divided by the invested capital, or

$$\text{rate of return} = \frac{\text{annual net profit}}{\text{invested capital}} \tag{12.4}$$

12.3 TIME VALUE OF MONEY

In making a decision to proceed with a project, the value of the equity capital must be analyzed. The company could lend it to someone else, or put it in a bank and receive interest. A decision to invest it in a new venture would be based, in part, on the rate of return being sufficiently high to warrant the chance of undertaking the venture. Nothing guarantees that the venture will be profitable. Certainly most businesses would want the rate of return to be higher than what they could receive by investing the funds and receiving interest. However important economic factors are, other factors influence a decision to proceed: for example, the continuance of a competitive company (not falling behind others in innovative and enhanced products); and social awareness (keeping your company's work force fully employed) are two other forces at play in determining a go/no-go decision.

Simple Interest

As the name implies, simple interest is the easiest to calculate and understand. A certain amount of money, the principal P, is borrowed for a stated time period. At the end of this time period the principal is paid back, plus the interest I that the lender receives for the use of the money. Let i be the rate of interest per period, n be the number of interest periods; the total interest is

$$I = Pni. \qquad\qquad 12.5$$

The entire amount F that must be paid back is the interest plus the principal, or

$$F = P + I = P + Pni = P\,(1 + ni). \qquad\qquad 12.6$$

E X A M P L E
12.1

A student wishes to borrow \$8,000 at 6 percent interest per annum with the entire amount to be paid back in five years. How much will this be?

SOLUTION:

$$F = 8,000[1 + 5(0.06)] = \$10,400$$

Thus, the student must repay \$10,400 after five years, to have the use of \$8,000 for that time period.

Compound Interest

Most often the lender requires intermediate payments on the loan. The borrower can use the interest that has been deferred; hence, the borrower is gaining additional capital and should pay interest on it. The total amount of money due at the end of the first interest period F is

$$F_1 = P + I_1 = P + Pi = P(1 + i),$$

and at the end of the second interest period, the amount due is

$$F_2 = F_1 + I_2 = F_1 + F_1 i = P(1 + i) + P(1 + i)i = P(1 + i)^2.$$

We may generalize for any n periods, and the amount due at the end of the nth period is

$$F_n = P(1 + i)^n, \qquad\qquad 12.7$$

where $(1 + i)^n$ is the compound interest factor.

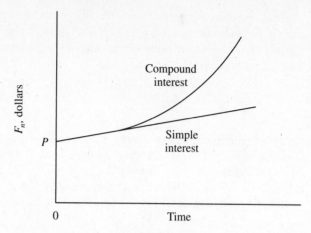

FIGURE 12.1 A graph of value versus time for compound and simple interest.

Referring to Example 12.1, if the principal were 6 percent compounded semiannually for five years, then $P = \$8,000$, $i = 3$ percent per interest period, and $n =$ ten interest periods. The total amount due after five years is

$$F_5 = 10,000(1 + 0.03)^{10} = \$10,751,$$

which includes $351 that must be paid due to compounding of the interest. Figure 12.1 illustrates the effect of compound and simple interest with time.

Equivalence

A dollar is not always equal to a dollar when time is involved. If funds can be put to productive use in a given year, then a dollar in the succeeding year should be equivalent to more than a dollar. For instance, a dollar of capital earning 6 percent interest is $1.06 at the end of one year. Thus, the original dollar is now worth 1.06 times as much. The effect of compounding accelerates this effect.

People often joke about the Dutch buying Manhattan Island for the equivalent of $24 in 1626. The joke is that people assume that $24 in 1626 is the same as $24 in 1992. Of course that is not true, which can be readily demonstrated by the equivalence concept. Let us assume that the alternative for the $24 was an investment at 5 percent interest for these intervening years. This does not include the effect of inflation or deflation, but gives an estimate of the tremendous value the money would have today.

$$F_{1992} = P_{1626}(1.05)^n = 24(1.05)^{366}$$

$$F_{1992} = \$1.37 \times 10^9 \text{ or } \$1.37 \text{ billion}$$

Not only was $24 close to the assessed value of the land at the time, the equivalent value of that money today is nothing to joke about.

Discounting

Equation 12.7 determines the value of the principal in the future; equally important is determining the worth or value of funds at the present time, when the funds will be received at some point in the future. For instance, you wish to invest P dollars at a certain interest, and the investment institution will pay you F dollars at the end of a certain period. Solving Equation 12.7 for P,

$$P = F_n /(1 + i)^n. \hspace{3cm} 12.8$$

Here P is the present worth of F_n. To obtain F_n funds in the future, P funds must be invested now.

E X A M P L E
12.2

Your aunt set up an investment account for you when you were five so that on your eighteenth birthday you would receive $8000. The interest on the account was 8 percent, compounded quarterly. Determine the amount she invested.

SOLUTION:

$$n = 13 \text{ years} \times 4 \text{ compoundings/year} = 52$$
$$i = 0.08/4 = 0.02$$
$$P = 8000\ [1/(1 + 0.02)^{52}]$$
$$P = 8000(0.3571) = \$2857$$

Your aunt invested $2857 13 years ago.

Annuities

Not many people can invest one large sum. More commonly a company, or person, makes a series of uniform payments to create a certain amount of capital. This is called an annuity and is graphed in Figure 12.2. An equal amount of funds, A, is invested at the end of each year for n years. At the end of the nth year the future equivalent of the sum of funds F_n is available for withdrawal.

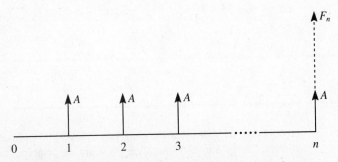

FIGURE 12.2 A cash flow diagram of the future worth F_n from investing n equal annuity payments.

We can derive an equation to determine the value of F_n as follows:

$$F_n = A + A(1 + i) + A(1 + i)^2 + \ldots + A(1 + i)^{n-1}. \qquad 12.9$$

     ↑   ↑   ↑

last   next to last   first

payment payment   payment

We have used Equation 12.8 for each yearly annuity payment with its compounded interest and summed them over the n years. Multiply both sides by $(1 + i)$,

$$(1 + i)F_n = A[(1 + i)^n - 1],$$

then subtract Equation 12.9 from this:

$$iF_n = A[(1 + i)^n - 1];$$

$$F_n = A\{[(1 + i)^n - 1]/i\}, \qquad 12.10$$

or

$$A = F_n\{i/[(1 + i)^n - 1]\}. \qquad 12.11$$

Both Equation 12.10 and 12.11 are useful, although the latter finds a greater use as we are most often interested in how much we should save periodically to accumulate a certain desired amount in the future.

E X A M P L E 12.3

A parent annually deposits $300 into a child's savings account that earns 7 percent annual interest. Determine the amount available on the child's eighteenth birthday.

SOLUTION:

Eighteen payments are made, the number of interest periods is 18, and the interest rate is 7.0 percent. From Equation 12.10 we find

$$F_{18} = 300 \{[(1 + 0.07)^{18} - 1]/0.07\} = \$10,199,$$

and of this $5,400 was from contributions and the remainder, $4,799, from interest.

E X A M P L E 12.4

A cooperative apartment building has a $500,000 balloon mortgage where the interest is paid annually and after ten years the principal must be paid. The coop board wants to determine how much money it should invest each year at 7.5 percent so it can pay off the mortgage at the end of ten years.

SOLUTION: Equation 12.11 provides the information that we need. $F_{10} = 500,000$, $i = .075$, and $n = 10$.

$$A = 500,000 \{0.075/[(1 + 0.075)^{10} - 1]\} = \$35,343$$

So \$35,343 must be invested annually to accumulate the \$500,000 after ten years.

■

Capital Recovery and Capital Recovery Factor

From Equation 12.8 we can determine the present worth of a future amount of money. If we combine this concept with that of annuities, then P_o may be thought of as the present worth of the future value of the annuity F_n and is related by the following equation and the cash flow diagram, Figure 12.3.

$$P_o = F_n /(1 + i)^n \qquad 12.12$$

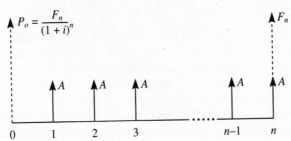

FIGURE 12.3 A cash flow diagram of the equivalent present worth P_o and the future worth F_n of n equal annuity payments.

This may be combined with Equation 12.10 to yield

$$P_o = A \left[\frac{(1 + i)^n - 1}{i}\right] \left[\frac{1}{(1 + i)^n}\right] = A \frac{(1 + i)^n - 1}{i(1 + i)^n}, \qquad 12.13$$

and solving for A yields

$$A = P_o \frac{i(1 + i)^n}{(1 + i)^n - 1}, \qquad 12.14$$

where the term in brackets is called the capital recovery factor.

■
E X A M P L E
12.5

A company borrows one million dollars for a capital improvement project. The loan is to be paid off over 12 years in equal monthly installments. The interest rate is 0.75 percent per month. Determine the monthly installment. Figure 12.4 is the cash flow diagram.

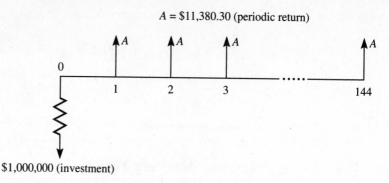

$$A = \$11,380.30 \text{ (periodic return)}$$

$1,000,000 (investment)

FIGURE 12.4 The cash flow diagram for Example 12.5.

SOLUTION:

The number of payment periods is $12 \times 12 = 144$, thus

$$A = 1,000,000 \left[\frac{0.0075(1.0075)^{144}}{(1.0075)^{144} - 1} \right] = \$11,380.30.$$

The monthly payment is $11,380.30, and the total amount paid over the 12-year period is $1,638,763. Let's extend this problem and determine the balance remaining after five years of payments have been made. Perhaps the company has an increase in earnings and might be able to reduce some of its debt.

Of the several ways to approach this problem, the quickest is to determine the present worth of the remaining payments after 60 payments. After five years there have been sixty payments, eighty-four payments remain, or from Equation 12.13,

$$P_{60} = 11,380.30 \left[\frac{(1.0075)^{84} - 1}{(0.0075)(1.0075)^{84}} \right] = \$707,330.76.$$

■

12.4 EVALUATION OF ALTERNATIVE PROPOSALS

Most frequently different proposals for equipment from different manufacturers will have different initial costs, operating costs, and useful life spans. The means to determine a common ground of comparison are surveyed in this section. A word of caution: we are assuming that we know what interest rates will be in the future, what the life expectancy of equipment is, and many other factors that really are not certain at all. A great deal of judgment must be exercised in actually determining these values, and often this involves probability and statistical analyses.

Present Worth

In calculating the present worth of an investment, with its estimated returns over a number of years, the entire set of cash flows, positive and negative, are projected to a single value at time zero. In determining whether or not to make an investment, a company establishes a minimum attractive rate of return (MARR) as a significant criterion in the decision process. The following examples will illustrate several aspects of calculating and interpreting present worth.

■
E X A M P L E
12.6

A company is considering an investment of $200,000 in new equipment that will provide income (revenues minus expenses) of $50,000 per year for five years. At the end of five years the equipment may be sold for $70,000. The company requires a MARR of 15 percent to undertake this project. Should it proceed?

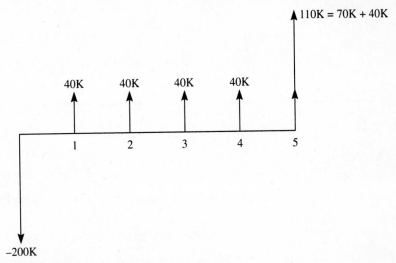

FIGURE 12.5 The cash flow diagram for Example 12.6.

SOLUTION:

Figure 12.5 is the cash flow diagram. Very simplistically, we note that the return is greater than the investment, but is it high enough? Equations 12.12 and 12.13 are used in calculating the present worth of the investment.

$$P_o = -200,000 + 50,000 \left[\frac{(1.15)^5 - 1}{(0.15)(1.15)^5} \right] + 70,000 \left[\frac{1}{(1.15)^5} \right]$$

$$P_o = -200,000 + 167,608 + 34,802 = +\$2,410$$

The present worth is positive, indicating that the MARR can be met. What happens, however, if the annual revenue falls short of the estimate by 10 percent? The return would be $45,000 annually, and the salvage value would be 10 percent less, or $63,000 after five years.

$$P_o = -200,000 + 45,000 \left[\frac{(1.15)^5 - 1}{(0.15)(1.15)^5} \right] + 63,000 \left[\frac{1}{(1.15)^5} \right]$$

$$P_o = -200,000 + 150,847 + 31,322 = -\$17,831$$

The negative sign indicates that the goal of a 15 percent MARR cannot be met under these circumstances. It also illustrates why projects initially must have a MARR higher than we might expect when comparing them to other investments, because greater uncertainty is associated with future events. ∎

It is possible to calculate the rate of return, i, from the equations for present worth. This calculation involves a trial-and-error procedure, where a value for i is assumed and the equation solved for the case of present worth equal to zero.

Future Worth

Just as the value of funds may be projected to time zero, they may also be projected to any time in the future.

∎
E X A M P L E
12.7

Calculate the future worth for the investment opportunity in Example 12.6.

SOLUTION: In this case Equations 12.7 and 12.10 are used.

$$F_5 = 70,000 + 50,000 \left[\frac{(1.15)^5 - 1}{0.15} \right] - 200,000 \left[(1.15)^5 \right]$$

$$F_5 = 70,000 + 337,119 - 402,271 = \$4,848$$

In this situation the future worth is greater than zero, indicating a future gain of $4,848 above the 15 percent return. In the second case, however,

$$F_5 = 63,000 + 45,000 \left[\frac{(1.15)^5 - 1}{0.15} \right] - 200,000 \left[(1.15)^5 \right],$$

and

$$F_5 = 63,000 + 303,407 - 402,271 = -\$35,864.$$

The future worth is negative, indicating that the company would have to pay an additional $35,864 to meet the 15 percent MARR, or that the goal of 15 percent would not be met. ∎

Annual Worth

Another measure of present worth, the annual worth, conceptually takes the present or future worth of an entire set of cash flows and converts it into an annuity.

■

E X A M P L E
12.8

Compute the annual worth of the investment given in Example 12.6.

SOLUTION:

The revenues are already in the form of an annuity, as they have an annual worth of $50,000. The initial investment of $200,000 must be converted to an annual amount, as must the salvage value of the equipment. This is called the equivalent annual cost (EAC) and may be most directly calculated by determining the present worth of the salvage value, adding it to the initial outlay, and then finding the annual annuity for this sum. There are other equations that can be derived based on this, but let's stick with this one presently.

$$P_o = -200,000 + \frac{70,000}{(1.15)^5} = -200,000 + 34,802 = -\$165,198$$

$$\text{EAC} = -165,198 \left[\frac{i(1 + i)^n}{(1 + i)^n - 1} \right] = -165,198 \left[\frac{0.15(1.15)^5}{(1.15)^5 - 1} \right] = -\$49,281$$

The annual worth AW is found by adding the annual revenue to the equivalent annual cost:

$$\text{AW} = 50,000 + (-49,281) = \$719.$$

The annual worth is greater than zero, indicating that the investment strategy is being met.

■

If a calculation of the annual worth is negative, as it would be for the second part of Example 12.6, the investment will not meet the target requirements. At times a negative annual worth must be accepted; it will not stop a project that is mandated by safety issues, for instance, and the task will be to find a solution with the least negative value.

12.5 ALTERNATIVES WITH DIFFERENT ECONOMIC LIVES

Present-worth analysis can be used as the comparative yardstick in evaluating different alternatives with the same economic life. For differing economic lives, annual worth is used as the yardstick of comparison. We have already discussed some of the factors involved in this type of decision process, such as investment costs, salvage value, and economic life. These values are somewhat arbitrary for some factors, such as economic life. The useful life of a piece of equipment is a matter of judgment and, in many cases, what the tax laws allow. The equipment is depreciated over that lifetime, with an assumed salvage value at the end. In general the shorter the economic life, the greater the salvage value.

We have also estimated the revenues that a new investment would provide. This may be easy to estimate for a new plant construction; the entire product line would have a certain value associated with it and would in turn produce a certain estimatable revenue. When the equipment to be purchased is only part of an entire assembly, the estimate of the revenue return is more difficult. The equipment may produce a savings over the equipment it replaces because of increased output, better efficiency. This savings may be thought of as revenue in calculating the annual worth.

Often the investment costs of one proposal are greater than that of another. Typically, the differences lie in reduced production costs associated with the higher priced alternative, as well as a higher salvage value. The costs are less usually because the machine requires less labor, either less-qualified or fewer people, and/or the machine is more efficient, consuming less electricity.

EXAMPLE
12.9

A company is planning to purchase a new machine to produce a product. The engineer in charge of the project has determined the costs associated with the two final alternatives.

	Machine X	Machine Y
Investment cost	$55,000	$72,000
Economic life	5 years	6 years
Annual revenue	$22,500	$23,500
Annual costs		
Labor	$ 6,200	$ 5,700
Electrical	$ 1,100	$ 1,200
Maintenance	$ 600	$ 650
Taxes, insurance	$ 600	$ 650
Total annual costs	$ 8,500	$ 8,200
Salvage value	$ 5,500	$ 7,000

The company will view positively investments with a MARR of 10 percent. Will these machines achieve that? Which is the better alternative?

SOLUTION: Calculate the annual worth of each machine.
Machine X:

$$P_o = -55,000 + [5500/(1.10)^5] = -51,585$$
$$\text{EAC} = -51,585 \{[0.10(1.10)^5]/[(1.10)^5 - 1]\} = -13,608$$
$$\text{AW} = 22,500 - 13,608 - 8,500 = \$392$$

Machine Y:

$$P_o = -72,000 + [7000/(1.10)^6] = -68,049$$
$$\text{EAC} = -68,049 + \{[0.10(1.10)^6]/[(1.10)^6 - 1]\} = -15,625$$
$$\text{AW} = 23,500 - 15,625 - 8,200 = -\$325$$

Thus, machine X meets the 10 percent MARR, whereas machine Y does not. It should be noted, however, that the economic analysis is one of the tools for decision making regarding *investment*. Strategic decisions will require more than economic analysis.

■

Depreciation

Whenever a company purchases an asset—a computer or any other piece of machinery or fixture—that item has an initial cost associated with it and an expected lifetime before it wears out. All machinery wears out; some becomes obsolete as well. At the end of the machinery's life it will have some salvage value, what the machinery can be sold for at that time. Thus, the total depreciation D for a piece of equipment is

$$D = P - SV,$$

where P is the purchase price and SV is the salvage value. Note that the price paid is used, not the value of the equipment. If you are able to purchase a computer system for $1500 instead of the list price of $2700, P is $1500. The price you paid is what must be depreciated, the initial value. If we assume that the machinery lasts n years, then the depreciation per year, d, is

$$d = \frac{P - SV}{n}.$$

This is straight-line depreciation: the machinery is assumed to decrease in value linearly with time.

Let P = $5000, SV = $500, and n = 5 years. The annual depreciation is

$$d = \frac{5000 - 500}{5} = \$900.$$

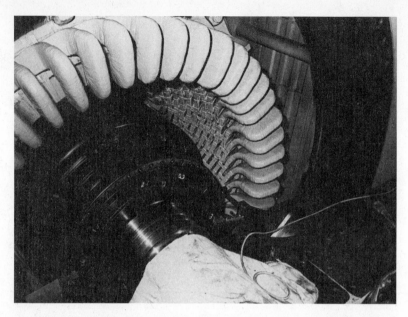

All equipment must be periodically inspected. This close view of an electric generator shows the rotor and windings. (Courtesy of Public Service Electric and Gas)

At the end of the first year the equipment would have a value of $5000 − $900 = $4100. This book value is what the item is valued at on the company's accounting ledger. In reality the equipment may have a resale value greater or smaller than this amount.

In practice, accelerated depreciation is used, but straight-line depreciation illustrates the concept well and is easier to use in calculations. Depreciation is one of the costs of doing business, with money being set aside from income to repurchase new equipment at the end of n years. When the equipment lasts longer than its projected life, the profits that had been claimed over the n depreciable years were understated, and the profits after the n years will be less than assumed, as the annual cost of depreciation will no longer be a business expense.

Breakeven Decisions

Competing machines may have not only different lifetimes and prices but also costs that vary as a function of the number of hours that the machines are used. Thus a machine may be a good choice if it is run 16 hours per day, but not if the operating time is 12 hours a day or less. Accordingly, some forecasting is required regarding the estimated production load that the machine will see over its useful life.

■
E X A M P L E
12.10

A manufacturing company is considering the purchase of two machines with the following costs.

	Machine X	Machine Y
Investment cost	$240,000	$320,000
Economic life	4 years	5 years
Annual maintenance	$ 10,000	$ 20,000
Operation cost per hour	$ 24	$ 16
Annual taxes and insurance	$ 12,000	$ 16,000
Salvage value	$ 50,000	$ 70,000

The company assumes that the cost of money will be 10 percent per year

SOLUTION:

No income is associated with the machines, so we must determine annual costs for each and find which is the least. The estimated annual cost (EAC) has three components: depreciation, interest expense (assume annual compounding), and annual expenditures.

$$EAC = d + I + AE$$

$$d = \frac{P - SV}{n}$$

$$I = \frac{i}{2}\left(\frac{n + 1}{n}\right)(P - SV) + iSV$$

$$AE = \text{all other costs}$$

Machine X:

$$d = (240,000 - 50,000)/4 = 47,500$$

$$I = \frac{0.10}{2}\left(\frac{5}{4}\right)(190,000) + (0.10)(50,000) = 16,875$$

$$AE = 10,000 + 12,000 + 24N,$$

where N is number of hours the machine operates annually.

$$(EAC)_x = 86,375 + 24N$$

Machine Y:

$$d = (320,000 - 70,000)/5 = 50,000$$

$$I = \frac{0.10}{2}\left(\frac{6}{5}\right)(250,000) + (0.10)(70,000) = 22,000$$

$$AE = 20,000 + 16,000 + 16N$$

$$(EAC)_y = 108,000 + 16N$$

Solve the equations for the case where N is the same for both machines. Setting the equations equal to each other yields a value of N of 2703 hours per year. If the machines are to operate more than 2703 hours per year, machine Y is the better choice, as the operating costs are less than those of machine X.

The significance of the number 2703 is more apparent if translated into hours per day. A plant may be in operation, not allowing for closings or holidays, 20 weekdays a month, 12 months a year, or 240 days a year. If the machine operates more than 11.25 hours a day, then machine Y is the better choice. If two shifts work at the plant, the machines would operate between 14 and 16 hours a day. Thus, a management decision is required: will the plant continue to work at this capacity for the next four to five years?

■

12.6 PUBLIC WORKS PROJECTS

Many public works projects require engineering economic analysis to prove their viability. However, viability and profitability are not the same. For public sector projects a different criterion is used. It is called the benefit-cost ratio, B/C, and the attempt is made to quantify the benefits and costs of a project to determine if a project is worthy of expenditure of public funds. This can be no small task, as we will see.

Battery energy storage systems, such as the one shown here, store electrical energy from power plants generated during times of low electric demand and release it during times of high electric demand. (Courtesy of Public Service Electric and Gas)

Very often the alternatives examined in the public sector have several benefits associated with them. For instance, the construction of a hydroelectric dam may generate electricity, provide a recreational environment, and afford flood control, all of which stimulate employment in a given geographic area. Assessing the values of the benefits is often controversial, as in flooding land to create a dam, or condemnation of buildings for a roadway. Different interest groups may claim conflicting costs and benefits.

The benefit-cost ratio is

$$\frac{B}{C + O + M},$$

where C is the capital costs, O is the operating costs, and M is the maintenance costs. Usually the analysis uses an equivalent annual worth or present worth to determine the costs.

E X A M P L E
12.11

A state wishes to develop its southwestern county by building a hydroelectric dam. The dam will serve a variety of purposes. It will generate electricity, which will be sold to the electric power grid or at the same price to a large industry seeking to relocate in the area; it will provide a lake for the citizens of the area and help a tourist industry; and it will provide water for irrigation of heretofore marginal farmland. The cost for the dam complex is $150 million, which includes construction and purchase of the land. In addition, the power generated is estimated to be 20 MW averaged over the year, and it can be sold for $0.06/kwh. Operating and maintenance costs associated with the facility are projected to be $1 million per year. Regional benefits associated with the recreational area are valued at $2 million, and the increased production from farmland is estimated to be $1 million. The funds for the development will be obtained by issuing a 30-year bond at 6 percent interest. Assume the payments are annual and include interest and principal. Determine the benefit-cost ratio.

SOLUTION:

The annual cost C_1 is found from Equation 12.14:

$$C_1 = 150,000,000 \left[\frac{(0.06)(1.06)^{30}}{(1.06)^{30} - 1} \right] = \$10,897,337.$$

The costs associated with operation and maintenance, C_2, are $1 million; hence the total cost C is $11,897,337.

The annual benefits are threefold; the first is from the electricity generated.

$$B_1 = (20,000 \text{ kW}) \left(\frac{24 \text{ hr}}{\text{day}} \right) \left(\frac{365 \text{ days}}{\text{year}} \right) \left(\frac{0.06 \text{ \$}}{\text{kwh}} \right) = \$10,512,000$$

The benefits associated with increased farmland utilization are $1 million, and from tourist development $2 million. Thus, the total benefit B is $13,512,000, and the benefit-cost ratio is

$$B/C = 13,512,000/11,887,337 = 1.13.$$

Since the benefit-cost ratio is greater than unity, the project is feasible. Whereas in this problem the benefits associated with improved farmland and tourist industry were simply assumed, in practice they are difficult to ascertain, and more difficult to persuade others who do not agree with the assumptions.

∎

REFERENCES

1. Kleinfeld, I.H. *Engineering and Managerial Economics*. Holt, Rinehart and Winston, New York, 1986.
2. Newman, D. *Engineering Economic Analysis*. 4th ed. Engineering Press, San Jose, CA, 1990.

PROBLEMS

12.1 You have earned $800 working during the summer and decide to invest the funds in a certificate of deposit (CD) that pays 8 percent interest per annum for a period of five years. How much will you receive at the end of the five years when cashing in the CD?

12.2 Plot a graph showing the effect of compound and simple interest (like Figure 12.1) for a principal of $10,000 and an annual interest rate of 8 percent for a period of ten years. Let the time interval be yearly.

12.3 You decide to start your own engineering firm. To obtain some start-up capital you borrow $10,000 from a relative and promise to repay it with 5 percent annual interest after four years. How much will you have to repay at that time?

12.4 When you graduate from college, age 22 years, your plan is to start an investment savings account, depositing $600 per year in the account, which you estimate should have an average annual interest of 6 percent over the next 40 years. How much money will the account have at the end of this time?

12.5 In Problem 12.4 let the payment be $50 per month and the account compounded monthly. Determine the final value in the account.

12.6 You wish to accumulate $5000 after five years by depositing X dollars annually at an 8 percent annual interest rate. Determine X.

12.7 One of the first things many students buy after they have their first job is a new automobile. Let's assume a new car costs $20,000 and you can take out a loan for 75 percent of its value at 12 percent per year for four years. What will the monthly payments be? As a freshman, what must you save per month, at 6 percent interest, compounded quarterly, to accumulate the down payment in four years?

12.8 Another purchase you anticipate is buying a house or condominium. Assume you anticipate that you will need an $80,000 mortgage, and the interest will be 12 percent per year, to be repaid in monthly payments over a 20-year period. Determine the monthly payments.

12.9 Determine the monthly payments and the total interest paid for $10,000 borrowed at 12 percent interest for the following time periods: 10, 15, 20, 25, and 30 years. What conclusions can you deduce from these data?

12.10 Write a computer program that will produce the monthly payments M for an amount X borrowed for N years at an annual interest rate i.

12.11 You have received $10,000 from your grandparents at the beginning of your college career. You decide to make quarterly withdrawals, leaving the balance in a savings account earning 8 percent interest, compounded quarterly as well. When you graduate in four years the account should be depleted. What are your withdrawals?

12.12 An engineer holds the patent on a certain device that others manufacture, paying him royalties. A manufacturing company offers to pay him $20,000 now or $8,000 next year, $10,000 in two years, and $5,000 in three years. The engineer estimates that the money can be invested at 8 percent compounded annually. Which offer should be accepted?

12.13 A company anticipates that it will spend $600 a year on maintenance of a given machine. Experience indicates that the maintenance cost will increase by $150 a year for the next ten years. Determine the present worth of the maintenance as well as the equivalent annual cost, assuming the interest is 12 percent compounded annually.

12.14 Your company borrowed $100,000 three years ago at an annual interest rate of 12 percent, to be repaid over five years. Two years ago it needed additional funding and borrowed $200,000 at 14 percent to be repaid over five years. Interest rates have declined to a current level of 10 percent, and the company wishes to consolidate the loans into one loan for five years. Determine the current monthly interest payments and the monthly interest payments with a consolidated loan. If the company can afford the current level of interest payments, what is the maximum loan it can receive?

12.15 The XYZ manufacturing company is considering the purchase of two robotic welding machines at a cost of $150,000 each. The life expectancy of the machines is six years, and the machines are expected to produce $60,000 net income per year. At the end of their expected life the machines can be sold for $40,000. The company wants to receive an MARR of 15 percent on its invested capital. Should it proceed? What is the actual MARR? (This will be a trial-and-error solution.)

12.16 Calculate the future worth and annual worth of the investment in Problem 12.15.

12.17 You have inherited $100,000 from a relative and wish to invest in real estate. You can purchase 200 acres of land and lease the land to a farmer for $2,000 per year. The land taxes are $1,200 and should remain constant. You expect to sell the land in five years for $200,000. What is the estimated rate of return on this investment, assuming 8 percent annual interest on accrued income?

12.18 A group of people form an investment company to purchase for $500,000 a commercial building that receives income of $80,000 per year from tenants. The tenants have signed ten-year leases, and the group feels confident that the tenants will stay in business for ten years. The maintenance costs and taxes associated with running the building are $20,000 per year and are expected to increase at 10 percent per year. After ten years the building may be sold for at least the purchase price. What is the minimum expected rate of return on this investment?

12.19 Parking is a problem on virtually all college campuses, with complaints from students, faculty, and staff. At Topnotch University the engineering students developed the cost analysis for the construction of a multistory parking garage. The university has no funds to build the structure, so the cost of the garage (with a 30-year life) must be derived from parking fees over a 15-year period. A MARR of 10 percent is considered reasonable and can be used as a basis for determining the size of the garage. The following is the construction cost, operating cost, and income for the garage. Determine the maximum number of levels that can be built and still meet the MARR criterion.

Garage levels	Total construction cost	Annual operating cost	Annual income
1	$ 750,000	$ 43,750	$ 125,000
2	2,750,000	75,000	437,500
3	4,500,000	100,000	712,500
4	6,000,000	118,750	1,012,500

12.20 As part of the modernization of a refinery it is necessary to construct a new pipeline and pumping station between two facilities. There are three different pipe sizes and pumping facilities that will satisfy the design constraints.

	Pipe diameter (mm)		
	350	400	450
Initial cost	$21,500	$30,000	$40,800
Annual cost	$ 7,680	$ 5,280	$ 3,360

The pipe will last ten years, after which time its value will be 10 percent of the initial cost. The cost of recovering the pipe is $2500, regardless of the pipe size. A MARR of 9 percent is desirable. Which pipe would you select?

12.21 You have been assigned the task of determining the costs of insulating and air conditioning a data-processing center. You found that the greater the insulation thickness, the less the air conditioning costs, for both initial and annual operating costs. The following are the data you collected:

	Case 1	Case 2	Case 3
Initial insulation cost	$40,000	$51,400	$68,570
Initial A/C cost	$59,400	$51,400	$43,400
Annual power cost	$ 7,400	$ 5,800	$ 4,700

The insulation has a 20-year life, whereas the air conditioning system has a 10-year life. Neither has any salvage value. Other annual costs are 5 percent of the initial costs. You are aware that the company policy for capital improvements is a MARR of 15 percent. You also wish to make a good impression and do more than is requested by performing an economic analysis of the above choices. Which case would you select?

12.22 As chief engineer for a small manufacturing company, you wish to modernize the present factory. The present equipment has been completely depreciated and paid for, but its productivity is less than that of new equipment. The following represents the pertinent data:

	Current	Proposed
Initial investment	none	$2.5 million
Estimated life	20 years	20 years
Annual operating cost	$560,000	$280,000
Interest	none	9%
Annual taxes and insurance	$62,500	$187,500
Annual maintenance	$375,000	$187,500

Prepare an annual cost comparison for the two cases.

12.23 For the dam project in Example 12.11, cost overruns were incurred in the construction of the project and now the initial capital cost is $225 million. On the bright side, the cost of electricity can be assumed to be $0.065/kwh. Determine the *B/C* ratio.

12.24 For Example 12.11 plot the *B/C* ratio for interest rates of 5, 6, 7, and 8 percent. At what interest rate is the ratio unity?

12.25 You have been appointed project engineer for a natural gas company. The company is planning the construction of a new pipe line and determined the following costs for different size pipes. The greater the pipe diameter, the greater its initial cost, but fewer pumping stations are required.

NATURAL GAS PIPE LINE DATA (DOLLARS IN THOUSANDS)

	Pipe diameter (mm)			
	200	250	300	350
Initial pipe cost	$10,500	$13,200	$15,600	$18,800
Number of pumping stations	10	7	4	2
Initial individual station cost	$ 350	$ 355	$ 385	$ 395
Annual pipe line maintenance	$ 300	$ 325	$ 410	$ 450
Total annual station maintenance	$ 355	$ 245	$ 115	$ 70
Annual pumping cost	$ 880	$ 500	$ 320	$ 160

The economic life of the stations and pipe line is 20 years, the MARR is 14 percent, and the annual taxes and insurance are 2.5 percent of the initial cost. Which configuration would you recommend?

COMPUTERS

C H A P T E R   O B J E C T I V E S

- To use the binary number system and understand its importance in digital computers.

- To construct logic diagrams and logic circuits.

- To follow the computer's processing cycle.

- To become familiar with common engineering computer applications.

- To learn about the evolution of computer programming languages.

(Photo courtesy of Milt and Joan Mann/
Cameramann International.)

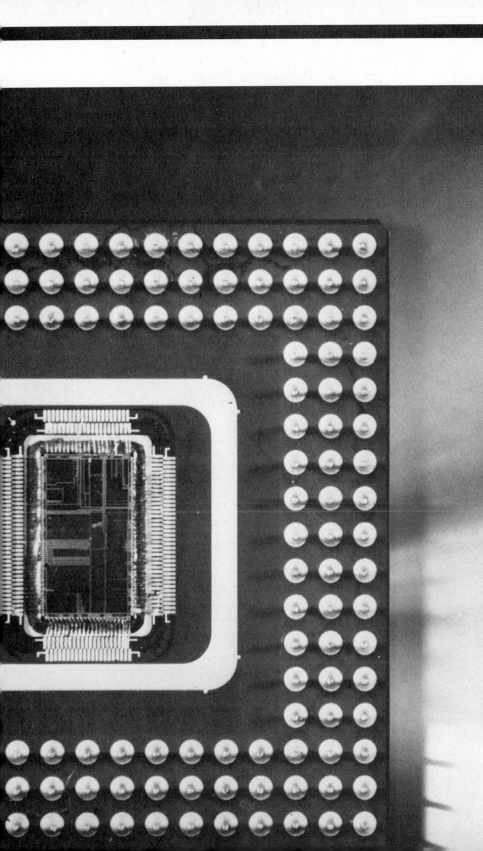

*M*uch of the technological world is becoming binary—digital computers and microprocessors are controlling an endless variety of devices from wristwatches to automobiles to satellites. The challenge for engineers is not only to advance the technological world but to interface it effectively with the very nondigital world of human discourse.

13.1 INTRODUCTION

Computers characterize our current society in much the same way that factories characterized the earlier industrial age. When one thinks about the industrial era, images of factories and smokestacks emerge. The postindustrial era, often called the computer era, is characterized by computers influencing many aspects of our lives, the types of services available to us, and the technologies we use. An engineer must be able to use computers as tools in solving problems in a variety of disciplines, and should understand conceptually, at least, how they function.

13.2 NUMBER SYSTEMS

Digital computers use binary numbers in performing the various calculational procedures. Information is stored in registers, devices that have two states that the computer reads as *on* or *off*; hence the binary system fits with the computer's functioning. Each digit is referred to as a bit (*bi*nary digi*t*), and a group of bits form a word. A word may contain only a certain number of bits, frequently eight, also known as a byte. In Figure 13.1 an eight-bit word is visualized as eight compartments containing one bit. The most significant bit is the one furthest to the left, whereas the least significant bit is furthest to the right, as indicated. The computer program that uses the words must determine the meaning of the bit location. The same word could mean a letter in the alphabet, a number with its sign, or instructions to the computer to perform a task. Because each location can have only an on or off value, it is adaptable to being magnetized or not. A magnetic recording head determines whether a bit location is magnetized (1 or 0), and this information is transferred for use in the computer.

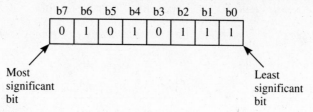

FIGURE 13.1 An eight-bit word, or byte.

All number systems have a certain commonality in their structure, and before we analyze the binary system, let's first examine the decimal, or base 10, system.

Consider the number 542.1. It is really a group of additions of numbers of various powers of ten:

$$
\begin{array}{ll}
500 & 5 \times 10^2 \\
40 & 4 \times 10^1 \\
2 & 2 \times 10^0 \\
+\ \ 0.1 & 1 \times 10^{-1} \\
\hline
542.1 &
\end{array}
$$

The sum of the right-hand column can be generalized with the coefficients of the various powers written as A_n; thus 5 is the coefficient of 10^2, and any number in decimal would have the decimal form of

$$A_n \times 10^n + A_{n-1} \times 10^{n-1} + \ldots + A_2 \times 10^2 + A_1 \times 10^1 + A_0 \times 10^0. \quad 13.1$$

The fractional terms are not included but are simply an extension of the exponent into the negative range.

Binary System

We can use the same concepts inherent in Equation 13.1 for any number system. Let the number system be the binary system, where the numbers are to the base 2. Thus, any number can be written as

$$B_n \times 2^n + B_{n-1} \times 2^{n-1} + \ldots + B_2 \times 2^2 + B_1 \times 2^1 + B_0 \times 2^0. \quad 13.2$$

In the decimal system A_n can assume any value between 0 and 9, and 10 is not an allowed coefficient. If we extend this to the binary system, B_n can assume any value between 0 and 1, and 2 is not an allowed coefficient.

Representations and Calculations

The following convention is used to distinguish among number systems. Let's write the numbers 2 and 10 in each of their number systems. In the decimal system 2 is 2×10^0, whereas in the binary system 2 is

$$(1 \times 2^1)\ (0 \times 2^0) = 10_2,$$

where the subscript 2 tells us that this is the base 2 system. The decimal system expresses the number 10 as

$$(1 \times 10^1) + (0 \times 10^0) = 10_{10},$$

and in the binary system it is expressed as

$$(1 \times 2^3) + (0 \times 2^2) + (1 \times 2^1) + (0 \times 2^0) = 1010_2$$
$$8 \quad + \quad 0 \quad + \quad 2 \quad + \quad 0 \quad = 10_{10}.$$

The conversion process relies on you remembering the values of powers of two.

E X A M P L E

13.1

SOLUTION:

Convert 1001110_2 into its decimal equivalent.

Count the number of places in the expression to determine how many terms are involved in the conversion. In this case there are seven.

$$(1 \times 2^6) + (0 \times 2^5) + (0 \times 2^4) + (1 \times 2^3) + (1 \times 2^2) + (1 \times 2^1) + (0 \times 2^0)$$

In decimal form,

$$64 + 0 + 0 + 8 + 4 + 2 + 0 = 78_{10}.$$

Octal and Hexadecimal Systems

Two other number systems are often used in computers, the octal, or base 8, system and the hexadecimal, or base 16, system. In the octal system the range of the coefficients is from 0 to 7. In the hexadecimal system a problem would occur if we let the coefficients vary from 0 to 15, as any value beyond 9 is not unique. To resolve this, the hexadecimal system uses 0 to 9 plus A to F to uniquely define the coefficients. Table 13.1 gives the equivalents for the number systems.

E X A M P L E

13.2

SOLUTION:

Convert 355 into octal and hexadecimal notation.

To convert to octal notation, we must remember the powers of eight.

$$(5 \times 8^2) + (4 \times 8^1) + (3 \times 8^0) = 543_8$$
$$320 \quad + \quad 32 \quad + \quad 3 \quad = 355_{10}$$

In hexadecimal notation we must use powers of 16.

$$(1 \times 16^2) + (6 \times 16^1) + (3 \times 16^0) = 163_{16}$$
$$256 \quad + \quad 96 \quad + \quad 3 \quad = 355_{10}$$

TABLE 13.1   DECIMAL, BINARY, OCTAL, AND HEXADECIMAL NUMBER EQUIVALENTS

Decimal	Binary	Octal	Hexadecimal
0	0	0	0
1	1	1	1
2	10	2	2
3	11	3	3
4	100	4	4
5	101	5	5
6	110	6	6
7	111	7	7
8	1000	10	8
9	1001	11	9
10	1010	12	A
11	1011	13	B
12	1100	14	C
13	1101	15	D
14	1110	16	E
15	1111	17	F
16	10000	20	10
17	10001	21	11
18	10010	22	12
19	10011	23	13
20	10100	24	14

Conversion between decimal and octal or hexadecimal is rare. Much more frequent, and simpler, is the conversion between the binary system and the octal or hexadecimal system. In this case the conversion is much simpler, brought about by eight and sixteen being powers of two.

E X A M P L E
13.3

Convert 10100110_2 into hexadecimal notation.

SOLUTION:

The first step is to partition the binary number into sets of four bits each. Zeros may be added as the most significant digits to create the groupings if necessary: 1010_2 and 0110_2.

The leftmost grouping is examined first.

$$1010_2 = (1 \times 2^3) + (0 \times 2^2) + (1 \times 2^1) + (0 \times 2^0) = 10_{10}$$

From Table 13.1 we know that $10_{10} = A_{16}$.

The rightmost grouping is

$$0110_2 = (1 \times 2^2) + (1 \times 2^1) + (0 \times 2^0) = 6_{10} = 6_{16}.$$

Combining the terms solves the problem.

$$10100110_2 = A6_{16}$$

Using Table 13.1 makes the conversion much easier than the previous example implies. Convert 100110101_2 into hexadecimal notation. Divide the term into groups of four, starting with the right-hand side and adding zeros as necessary to the left-hand side to create the necessary complete groupings. Thus, the three groups are 0001_2, 0011_2, and 0101_2. These are 1_{16}, 3_{16}, and 5_{16}, or 135_{16}.

13.3 MATHEMATICAL OPERATIONS

Operations that are easy for us in the decimal system become more complicated in the binary system. In part it is because our mindset is conditioned to the decimal world.

Addition

Consider the following addition operations in binary: $0 + 0 = 0$; $0 + 1 = 1$; $1 + 1 = 0 = 10_2$. In the last example the one is carried to the next place, leaving a zero in the original location. To help us with additions we can use a truth table, as shown in Table 13.2.

TABLE 13.2 THE BINARY ADDITION TRUTH TABLE

Augend	Addend	Sum	Carry
0	0	0	0
0	1	1	0
1	0	1	0
1	1	0	1

It helps to remember terms from arithmetic; *augend* is a number to which another is added, while *addend* is a number that is added to another. The result is a sum and/or a carry to the next place. The table represents all the possible combinations that can occur. When performing an addition caution is required when carrying a one, as it may cause an additional carry.

For instance, add the binary numbers 1 and 1010 as follows.

$$\begin{array}{r} 1010 \\ \underline{0001} \\ 1011 \end{array}$$

If the binary numbers 1 and 1111 are added, we get the following results.

$$\begin{array}{r} 1111 \\ +\quad 0001 \\ \hline 10000 \end{array}$$

The carry created zeros and an additional place was required.

Subtraction

Subtracting binary numbers produces another truth table (Table 13.3). *Subtrahend* is the number to be subtracted, *minuend* is the number from which the subtrahend is subtracted, yielding a *difference* between them. At times it is necessary to borrow from the next place.

TABLE 13.3 THE BINARY SUBTRACTION TRUTH TABLE

Minuend	Subtrahend	Difference	Borrow
0	0	0	0
0	1	1	1
1	0	1	0
1	1	0	0

Subtract the following binary numbers. The italic one over the minuend indicates that a borrow was necessary.

$$
\begin{array}{r}
\textit{1} \\
11101 \\
-\ 1011 \\
\hline
10010
\end{array}
\qquad
\begin{array}{r}
\textit{1111} \\
10000 \\
-\ \ \ 11 \\
\hline
1101
\end{array}
$$

Actually, computers subtract by creating a complement of the subtrahend and adding this to the minuend. Thus, both addition and subtraction become processes of addition. We will see that multiplication and, by extension, division also become processes in addition that the computer performs very quickly.

Multiplication

Multiplication is an addition process; if we are multiplying 18×12, for instance, 18 is added to zero 12 times or to itself 11 times. In the process of multiplication in the decimal system, we first multiply 18×2 and add this to the multiplication of 18×10. In the binary system a similar process occurs. First, the binary system multiplication table is as follows:

$$
\begin{aligned}
0 \times 0 &= 0 \\
0 \times 1 &= 0 \\
1 \times 0 &= 0 \\
1 \times 1 &= 1
\end{aligned}
$$

Let's convert 18×12 into its binary equivalent of 10010×1100.

$$
\begin{array}{r}
10010 \\
\times\ 1100 \\
\hline
00000 \\
00000 \\
10010 \\
10010 \\
\hline
11011000 = 216_{10}
\end{array}
$$

In a similar manner division may be viewed as repeated subtractions. In a computer subtractions are actually performed as additions of the complement, and this is the process that is used for division as well.

Binary-coded Decimal

An amalgam of the binary and decimal number systems is the binary-coded decimal (BCD) system. In this system the decimal number, such as 2754, is converted into its binary form by converting the constituent numbers individually. Each decimal number is represented by a four-digit binary number.

$$
\begin{array}{cccc}
2 & 7 & 5 & 4 & \text{Decimal} \\
0010 & 0111 & 0101 & 0100 & \text{BCD}
\end{array}
$$

13.4 LOGIC DIAGRAMS

Logic diagrams illustrate the path of information within a computer. Computer engineers design logic diagrams using three fundamental elements, the AND gate, the OR gate, and the inverter. Other circuit elements may be derived from these. The gates may be constructed physically from a wide variety of electrical/electronic switches, diodes, transistors, or fluidic devices. Whereas the same logic diagram can be used regardless of the physical device, the computer engineer must judge the speed of signal transmission, cost, and availability of the device in specifying the actual circuit construction. In computers the gates are transistors, contained within an integrated circuit (IC) chip.

AND Gate

The AND gate is a device whose output is a logic 1 only if both inputs are a logic 1. If one input is a logic 0, then the output will be a logic 0. Logic 1 and 0 states are, by convention, known as closed and open, high and low, or true and false.

Consider the simple electric circuit shown in Figure 13.2a. A battery is connected to a light, which may be on only if both switches A and B are closed.

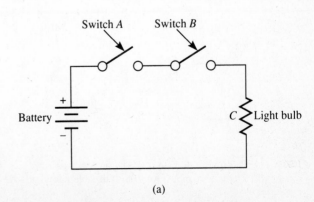

FIGURE 13.2 The logic AND gate. (a) The analogous electric circuit.

(a)

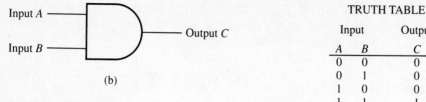

TRUTH TABLE

Input		Output
A	B	C
0	0	0
0	1	0
1	0	0
1	1	1

Logic equation:

$C = A \cdot B$

(b)

(c)

FIGURE 13.2 (b) The gate symbol. (c) The truth table and logic equation.

If either one is closed without the other, no current can flow, and the light will be off. Figure 13.2b symbolizes this AND gate. The truth table and the logic equation for the AND gate is shown in Figure 13.2c. The table reflects the only way for the light to be on, the logic 1 output state. The Boolean logic equation, $C = A \cdot B$ or $C = AB$, is read "C equals A and B," meaning that for C to be logic 1, both A and B must be logic 1.

OR Gate

The electric circuit analogy for the OR gate is shown in Figure 13.3 with the accompanying OR gate symbol and the truth table. In this situation the light will be on (logic 1 state) if either or both switches are closed. The truth table reflects these conditions. The logic equation is read "C equals A or B." Notice that the symbol + is the Boolean symbol for OR, not addition.

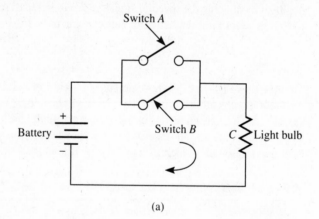

(a)

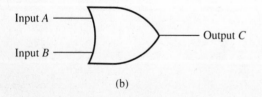

Output C

(b)

TRUTH TABLE

Input		Output
A	B	C
0	0	0
0	1	1
1	0	1
1	1	1

Logic equation:

$C = A + B$

(c)

FIGURE 13.3 The logic OR gate. (a) The analogous electric circuit. (b) The gate symbol. (c) The truth table and logic equation.

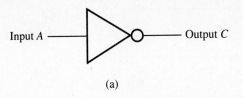

TRUTH TABLE		Logic equation:
Input	Output	$C = \overline{A}$
0	1	
1	0	

(a)   (b)

FIGURE 13.4 The logic inverter. (a) The symbol. (b) The truth table and logic equation.

Inverter

The last logic gate we will consider is the inverter, or NOT gate. It simply changes the input state from a logic 1 to a logic 0 or vice versa. This is shown in Figure 13.4a, with the truth table and logic equation in Figure 13.4b. The equation is read as "C equals A not" or "C equals not A."

13.5 LOGIC CIRCUITS

It is possible to join two or more gates to provide a variety of logic functions. It is not necessary to be limited to two input signals; any number of inputs may be considered. We will consider just three, as illustrated schematically in Figure 13.5. For a logic 1 state as the output, we need logic 1 states for the inputs to gate 2, and this is achieved only if logic 1 states are inputs to gate 1.

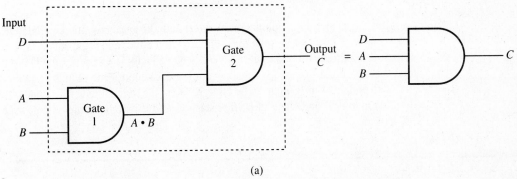

(a)

TRUTH TABLE				Logic equation:
Input			Output	$C = D \cdot (A \cdot B)$
D	A	B	C	
0	0	0	0	
0	0	1	0	
0	1	0	0	
0	1	1	0	
1	0	0	0	
1	0	1	0	
1	1	0	0	
1	1	1	1	

FIGURE 13.5 A three-input AND gate circuit. (a) Schematic diagram and symbolic representation. (b) The truth table and logic equation.

(b)

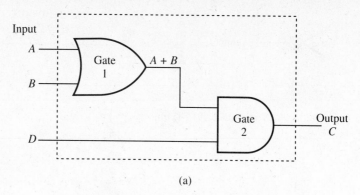

(a)

TRUTH TABLE

Input			Output
A	B	D	C
0	0	0	0
0	0	1	0
0	1	0	0
0	1	1	1
1	0	0	0
1	0	1	1
1	1	0	0
1	1	1	1

Logic equation:

$$C = (A + B) \cdot D$$

(b)

FIGURE 13.6 A three-input AND/OR gate circuit. (a) Schematic diagram. (b) The truth table and logic equation.

Let's consider the combination of AND and OR gates creating an AND/OR gate. Figure 13.6 illustrates the logic circuit, truth table, and logic equation. For a logic 1 condition at the output of gate 2, we need a logic 1 state for the D input and for the output of gate 1. To achieve a logic 1 state for the output of gate 1, we need neither A or B inputs to be logic 1. Thus, the truth table verifies that with input D in logic 1 and either or both of states A and B in logic 1, the output, state C, is logic 1.

E X A M P L E

13.4

You have been given the task of designing the logic circuit for the control of a traffic light. It is desired that the light turn green when approaching cars are sensed and when the preceding traffic light is green. Thus, if a car is approaching and if the preceding traffic light is green, then turn this traffic light green.

SOLUTION:

These are the requirements of an AND gate, so the equation is $C = AB$, where C stands for turning on the traffic light, B stands for the previous light being green, and A stands for cars approaching. A schematic is shown in Figure 13.7.

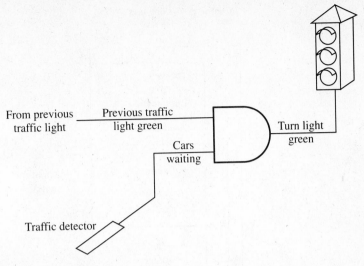

FIGURE 13.7 A traffic-light control circuit.

13.6 COMPUTER PROCESSING CYCLE

Computers perform simple operations with tremendous speed and accuracy. Information enters the computer through an input device, such as the keyboard at a computer terminal. The computer performs various calculations with these data in the central processing unit (CPU), and the results are passed to an output unit, such as a printer or video display. Thus, for an engineering report you input the keystrokes for the various words, the CPU translates these into binary numbers and then back into letters, which are displayed on the video terminal. For another part of the report you may input experimental data; the CPU will perform calculations using the data and send the results of the calculations to the printer.

Figure 13.8 illustrates the various components in a computer system. The input and output devices include tape drives, keyboards, disk drives, printers, card readers, card punches, video terminals, and speakers. The heart of the computer is the CPU. It consists of three units, the processing unit, more typically called the arithmetic logic unit (ALU), the main storage unit, and the control unit.

The ALU performs all the arithmetical calculations and makes logical decisions as well, such as whether or not a number is greater than the previous one. How does the ALU receive the information in the first place? When you input the data the control unit directs this information to the storage unit so it is available when the ALU needs it for calculations. The storage unit then holds the results of the calculations for the output unit. All these functions can take place on the surface of a microchip.

Central processing unit (CPU)

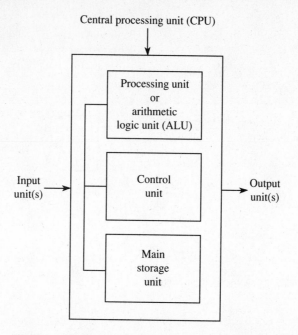

Input
unit(s)

Processing unit
or
arithmetic
logic unit (ALU)

Control
unit

Main
storage
unit

Output
unit(s)

FIGURE 13.8 The basic components in a computer system.

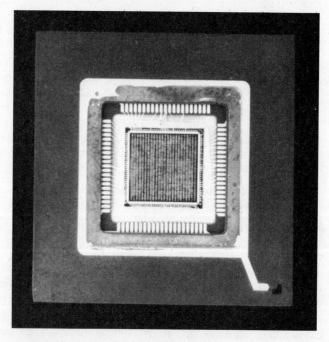

A computer chip. (Courtesy of Grumman Corporation)

The ALU performs all the additions, subtractions, multiplications, and divisions as well as logical decision making. The computer decision-making capabilities are built-in, or rather programmed in machine language instructions, and do not require additional programming by the user. Three of the typical comparative operations are the "equal to," "greater than," and "less than" operations.

How do you determine if an item is stocked in a warehouse's inventory? Enter the item's identification code, and the computer searches for a number equal to this; if found, then the item is in the inventory. Credit cards are frequently used to purchase items in a store. As part of the charging process, the store will check your card number with a central computer to determine if the available balance is greater than the amount you wish to charge. The "less than" function can be used in situations requiring that a certain minimum quantity be met, such as in a temperature-sensing system.

How does the computer know how to process the data? The set of instructions directing this is in the computer program. The programs are held in main storage, enabling the computer to execute the instructions when necessary, using the data held in storage. The computer must keep track of where the data are located and has memory locations that designate where a piece of data is located, its address, and what its value is. Figure 13.9 diagrams the two aspects of the memory location.

As a computer user, the engineer will most frequently be running software to solve a problem. He or she will be interfacing with the applications program, a CAD program for instance. This program must eventually use the computer hardware, the electronic bits, to perform the various operations required. To accomplish this, the program interfaces with the computer's operating system, such as MS-DOS, UNIX or CP/M. These operating systems control the operation of the computer hardware. They manage the hardware, the data, and the software programs. The operating system also directs the input and output devices, allows several users to access the same computer, and allows one person to run several programs simultaneously.

A microprocessor is a CPU in miniature form, fabricated on an IC chip encased in plastic for strength and for holding the connecting pins. Printed circuit boards comprise many IC chips that may include a microprocessor. As you are

Address	Value
0	−16.21
1	0.1695
2	1000.1
3	−24
•	•
•	•
•	•

FIGURE 13.9 Computer memory location elements.

A microprocessor is contained on a single silicon chip, small enough to fit on a fingertip. (Courtesy of ASME)

undoubtedly aware, circuit boards are used in a variety of applications, from automobile control systems to children's toys.

Microcomputers commonly have two types of memories, ROM (read-only memory) and RAM (random access memory). Once they are fabricated, ROM memories cannot be altered, but neither can their information be lost, and it is instantly available when the power is turned on. They are not affected by loss of power as RAM memories are. They are found in controllers, such as those used in microwave ovens, and in calculator functions. Some of the disadvantages of ROM are that the memory cannot be used for other purposes, it cannot be written into, and it is not easily programmed except at the factory, where the programming is performed on a large volume basis. One type of ROM can be programmed by the purchaser, but it remains permanent after it is programmed. This is called PROM (programmable read-only memory). Not to be defeated by a PROM that couldn't be reprogrammed, electrical engineers created an EPROM (erasable programmable read-only memory), which can be reprogrammed using special procedures, such as erasing the EPROM with ultraviolet light and then reprogramming the chip.

A common place for ROMs is in calculators for use by special function keys, such as the exponential function. The function may be written as an infinite series, such as

$$e^x = 1 + x + \frac{x^2}{2!} + \frac{x^3}{3!} + \frac{x^4}{4!} + \ldots$$

When the key is pushed, the exponential is calculated using the ROM program and the result displayed. This explains the delay from the start of the execution until the number is displayed on the calculator.

RAM memories are programmable, a singular advantage, but they are susceptible to power loss, a singular disadvantage. The program size that a microcomputer can execute depends on the RAM available. Microcomputers will have both RAM and ROM memories to serve different functions.

13.7 COMPUTER APPLICATIONS

Perhaps the area of greatest impact that the computer has had is on the wide-ranging applications we can now perform with it. Imagine a circle and smaller circles around its perimeter and tangent to it. The inner circle represents a computer, and the outer circles represent applications using the computer, such as word processing, computer-aided design, computer-aided manufacturing, spreadsheets, data-base management, desktop publishing, finite element analysis, and robotics. Each of the applications is linked to the others through the computer. As the power and utility of the computer increases, this "diameter" increases, and new circles of applications fit on the perimeter.

The computer affects our lives and work in myriad ways, perhaps none more fundamentally important than the way goods are manufactured. Before the advent of the industrial age, people handcrafted devices; they literally made them. With the industrial revolution, machines made the devices; people operated the machines. Now people supervise a computer, which operates a

The X-29 is a forward swept wing experimental aircraft. The aircraft requires computer control, as the human response time of the pilot is not quick enough to guide the plane. The forward swept wings allow better performance at supersonic speeds than the traditional rear swept wing design. (Courtesy of Grumman Corporation)

machine, which makes the device. This occurs daily as computers operate automated bank tellers, robots, and numerically controlled machines.

The role of people in the computer age is that of supervisor, contrasted with operation in the earlier industrial age. As such, engineers use a variety of software packages to supervise the computer's operation and need to know how information is transmitted between programs and within programs for proper supervision. Several codes have been established for this communication. The most popular is an alphanumeric code called ASCII (American Standard Code for Information Interchange), a seven-bit code that allows 128 (2^7) code combinations for letters, numbers, and symbols. Table 13.4 lists the binary

TABLE 13.4 ASCII CODE CHARACTERS AND THEIR BIT EQUIVALENTS

Character	ASCII code							Character	ASCII code							Character	ASCII code						
space	0	1	0	0	0	0	0	@	1	0	0	0	0	0	0		1	1	0	0	0	0	0
!	0	1	0	0	0	0	1	A	1	0	0	0	0	0	1	a	1	1	0	0	0	0	1
"	0	1	0	0	0	1	0	B	1	0	0	0	0	1	0	b	1	1	0	0	0	1	0
#	0	1	0	0	0	1	1	C	1	0	0	0	0	1	1	c	1	1	0	0	0	1	1
$	0	1	0	0	1	0	0	D	1	0	0	0	1	0	0	d	1	1	0	0	1	0	0
%	0	1	0	0	1	0	1	E	1	0	0	0	1	0	1	e	1	1	0	0	1	0	1
&	0	1	0	0	1	1	0	F	1	0	0	0	1	1	0	f	1	1	0	0	1	1	0
'	0	1	0	0	1	1	1	G	1	0	0	0	1	1	1	g	1	1	0	0	1	1	1
(	0	1	0	1	0	0	0	H	1	0	0	1	0	0	0	h	1	1	0	1	0	0	0
)	0	1	0	1	0	0	1	I	1	0	0	1	0	0	1	i	1	1	0	1	0	0	1
*	0	1	0	1	0	1	0	J	1	0	0	1	0	1	0	j	1	1	0	1	0	1	0
+	0	1	0	1	0	1	1	K	1	0	0	1	0	1	1	k	1	1	0	1	0	1	1
,	0	1	0	1	1	0	0	L	1	0	0	1	1	0	0	l	1	1	0	1	1	0	0
−	0	1	0	1	1	0	1	M	1	0	0	1	1	0	1	m	1	1	0	1	1	0	1
.	0	1	0	1	1	1	0	N	1	0	0	1	1	1	0	n	1	1	0	1	1	1	0
/	0	1	0	1	1	1	1	O	1	0	0	1	1	1	1	o	1	1	0	1	1	1	1
0	0	1	1	0	0	0	0	P	1	0	1	0	0	0	0	p	1	1	1	0	0	0	0
1	0	1	1	0	0	0	1	Q	1	0	1	0	0	0	1	q	1	1	1	0	0	0	1
2	0	1	1	0	0	1	0	R	1	0	1	0	0	1	0	r	1	1	1	0	0	1	0
3	0	1	1	0	0	1	1	S	1	0	1	0	0	1	1	s	1	1	1	0	0	1	1
4	0	1	1	0	1	0	0	T	1	0	1	0	1	0	0	t	1	1	1	0	1	0	0
5	0	1	1	0	1	0	1	U	1	0	1	0	1	0	1	u	1	1	1	0	1	0	1
6	0	1	1	0	1	1	0	V	1	0	1	0	1	1	0	v	1	1	1	0	1	1	0
7	0	1	1	0	1	1	1	W	1	0	1	0	1	1	1	w	1	1	1	0	1	1	1
8	0	1	1	1	0	0	0	X	1	0	1	1	0	0	0	x	1	1	1	1	0	0	0
9	0	1	1	1	0	0	1	Y	1	0	1	1	0	0	1	y	1	1	1	1	0	0	1
:	0	1	1	1	0	1	0	Z	1	0	1	1	0	1	0	z	1	1	1	1	0	1	0
;	0	1	1	1	0	1	1	[	1	0	1	1	0	1	1	{	1	1	1	1	0	1	1
<	0	1	1	1	1	0	0	\	1	0	1	1	1	0	0	|	1	1	1	1	1	0	0
=	0	1	1	1	1	0	1	]	1	0	1	1	1	0	1	}	1	1	1	1	1	0	1
>	0	1	1	1	1	1	0	^	1	0	1	1	1	1	0	~	1	1	1	1	1	1	0
?	0	1	1	1	1	1	1	—	1	0	1	1	1	1	1	delete	1	1	1	1	1	1	1

equivalents of ASCII codes. For instance, $S = 1010011$, while $s = 1110011$. Data files of ASCII code can be created in one applications program and transferred to another applications program, where these data can be used for another function. This commonly occurs in finite element analysis, where the graphics are created in a CAD system and the coordinates are placed in data files and loaded into a finite element program for analysis.

Word Processing

Word processing is an odd term for software that eases the tedium of writing. Computers were first created for data manipulation, or processing; later, as more sophisticated software was created, words could also be manipulated, or processed—hence the terminology. The word processing software that is available today is quite sophisticated and easy to use. It can perform a variety of functions for you, but you must be the creative genius.

Word processing software is menu-driven and can be used almost immediately, as the screen on which you see the typed copy has indexes to the help screens. Not only can you move quickly through the text and move chunks of text from one location to another, you can replace words easily. Many programs have a text replace feature, so a word can be replaced throughout the text, useful for correcting a misspelling. The software often includes a dictionary that will check the spelling of the text for you.

You can also import text from another document, so you do not have to rewrite repeated material. In addition you can determine how the output copy should look—words underlined, or double-struck for boldness.

For you as a student, word processing software is very useful in all your courses, from English to engineering. The papers and reports that you write have the potential to reflect very well upon you. In engineering practice you will be communicating frequently, and word processing is an essential part of your job in report and proposal writing.

Spreadsheets

Spreadsheets are a computer application that has gained tremendous popularity with the accounting and financial world and is now gaining popularity with the engineering community. A spreadsheet is a computer program that analyzes data. The first personal computer spreadsheet, VisiCalc, was developed by Dan Bricklin and Bob Franklin to solve financial planning problems. The many similar programs today include Lotus 1-2-3 and Excel.

Data analysis involves separating interrelated information into constituent parts, then varying the parts to determine the effect on the whole. The data must be separable and arranged in columns and rows. In financial planning and accounting, a vast array of numbers must be manipulated, such as in the grid illustrated in Figure 13.10. The intersection of a row and column is a *cell*. A cell

```
E19:
```

```
      ......A/.......B/.......C/.......D/.......E/......F/......G/.......H/
 1             OPTIMUM WIDGET PRICING STRATEGY
 2  LIST       NO. OF       DISC.    NO. OF       TOTAL
 3  PRICE      PURCHASES    PRICE    PURCHASES    SALES
 4         5   8000            3      3000        49000
 5      5.15   7850         3.09      2950        49543
 6       5.3   7700         3.18      2900        50032
 7      5.45   7550         3.27      2850        50467
 8       5.6   7400         3.36      2800        50848
 9      5.75   7250         3.45      2750        51175
10       5.9   7100         3.54      2700        51448
11      6.05   6950         3.63      2650        51667
12       6.2   6800         3.72      2600        51831.99
13  6.349999   6650         3.81      2550        51943
14  6.499999   6500          3.9      2500        51999.99
15  6.649999   6350         3.99      2450        52002.99
16  6.799999   6200         4.08      2400        51951.99
17  6.949999   6050     4.169999      2350        51846.99
18  7.099999   5900     4.259999      2300        51687.99
19
20
```

FIGURE 13.10 The output of a spreadsheet.

may contain a label, value, or formula. A label is a set of characters describing a column or row. In most spreadsheets when you start a cell entry with a letter, the program assumes you are entering a label. Values must start with a number, and a formula with a plus sign, alerting the program that an equation is being entered. (Some spreadsheet programs use other designators, such as an equal sign.)

When you first encounter a spreadsheet, the screen will indicate the columns A, B, C, . . . , and the rows 1, 2, 3, . . . , and the cell A1 will be highlighted, as illustrated in Figure 13.11.

```
A1:
```

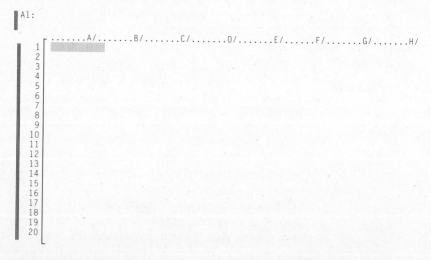

```
      ......A/.......B/.......C/.......D/.......E/......F/......G/.......H/
 1  ▓▓▓▓▓▓▓▓▓▓▓
 2
 3
 4
 5
 6
 7
 8
 9
10
11
12
13
14
15
16
17
18
19
20
```

FIGURE 13.11 The display of a spreadsheet with cell A1 highlighted.

EXAMPLE
13.5

You and a friend have decided to go into business manufacturing and selling widgets. Being a cautious person, however, you wish to determine if the business can sell the widgets at a price sufficiently high to cover manufacturing costs and provide a reasonable rate of return to the business. To that end you have performed a market survey and determined that if a widget were priced at $5.00 a unit, you could expect annual direct sales of 8000 units. Furthermore, if you discount the price by 40 percent, you can obtain additional sales of 3000 units to a store selling the items under their brand name. The marketing analysis further reveals that for every $0.15 increase in list price, direct sales will decrease by 150 annually and discounted sales by 50. Create a spreadsheet that will help you establish the initial list price for maximum total sales revenue.

SOLUTION:

With the cursor at cell A1, type in the heading "OPTIMUM WIDGET PRICING STRATEGY." Return the cursor to cell A2, and type in the second line with adequate spaces for the columns as shown. Repeat for cell A3. With the cursor indicating cell A4, type in 5.00, move to B4 and type in 8000. Move the cursor to C4. You could calculate the discounted value, but have the program perform this for you. This will allow you to change values later for a sensitivity analysis. Since you are typing in an equation, enter +0.6∗A4; this will appear in the edit cell in the far upper left corner, and the program will automatically calculate the value, 3.00, which will appear in cell C4. For D4 enter the sales estimate. For

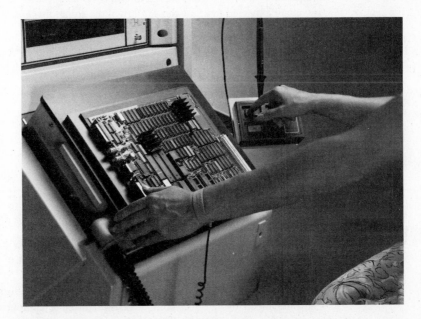

The reliability of computers is assured by testing of the many components before assembly. This device is a vacuum-assisted circuit board tester. (Courtesy of Hewlett-Packard)

cell E4 you wish total sales revenue to appear; thus enter an equation by typing +A4*B4+C4*D4. The spreadsheet will multiply the first two cells and add them to the value of the product of the third and fourth cells. With cell A5 you do not enter $5.15, but +A4+0.15; for cell B5, +B4−150; for C5, +0.6*A5; for D5, +D4−50; and for E5, +A5*B5+C5*D5. This must be repeated for each cell throughout the program. Obviously, this is a tedious process for one iteration, but the advantage will appear later. If you perform this input, Figure 13.10 will result. In this situation the list price for maximum revenue is $6.65. However, this is based on your marketing analysis of what the purchases will be and how they will change with changing prices.

Another advantage of the spreadsheet is its use in a sensitivity analysis. Vary the initial price and sales projections to determine the effect on total revenue. In this example, if the initial price is increased to $6.00, all other variables remaining unchanged, the maximum revenue will be $60,207 at a list price of $7.05. If the sales projections are varied with the list price remaining at $5.00, the following results: for 9,000 purchases at a $5.00 initial price, a maximum revenue of $58,797 occurs at a list price of $6.95; for 7,000 purchases at a $5.00 initial price, a maximum revenue of $45,632 occurs at a list price of $6.05.

■

For commercial applications the cells can number in the thousands, and the analysis can be quite complex. You can see the importance of the underlying equations in determining the maximum revenue point, in this case. As engineers you are expected to know the validity of the equations used, with their associated assumptions.

If spreadsheets were used only in financial analyses, their application in engineering courses would not be very great. However, they may also be used to solve equations and sets of linear equations.

Consider the equation

$$(16 + 2z/x^3)(y/z^2).$$

Solve the equation for values of $x = -2$, $y = 4.5$, and $z = 3$. Let x be the value of cell A1, y the value of cell B1, and z the value of cell C1. For cell D1 write the formula, +(16.0+(2*C1)/A1^3)*(B1/C1^2). The result is 7.625, and the screen is shown in Figure 13.12.

Computer-Aided Design

The acronym CAD stands for computer-aided design. It most often appears in conjunction with CAM, computer-aided manufacturing. CAD/CAM refers not to one activity in the design or manufacturing process, but to many activities that are enhanced by using a computer, most often a computer workstation. The workstation is a very powerful computer that has accessories, such as display and input/output devices, and is connected with a mainframe computer. An engineer at the workstation can call up programs from the mainframe and use them in

```
D1:  (16+2*C1/A1^3)*(B1/C1^2)

       .......A/.......B/.......C/.......D/.......E/......F/.......G/.......H/
     1       -2      4.5        3     7.625
     2
     3
     4
     5
     6
     7
     8
     9
    10
    11
    12
    13
    14
    15
    16
    17
    18
    19
    20
```

FIGURE 13.12 The spreadsheet display of the equation $(16 + 2z/x^3)(y/z^2)$ at $x = -2$, $y = 4.5$, and $z = 3$. The result is in cell D1.

performing a design or analysis. The results can be seen at the workstation as well as communicated back to the mainframe. This same information can be transmitted from the mainframe to the computers used in the manufacturing process. CAD lets the engineer perform creative and multiple designs, as the computer does the necessary repetitive calculations.

Because of the workstation's computing power, memory, and storage capabilities, an engineer can see the results of a design and/or analysis on the CRT (cathode ray tube) display. The graphics display is most often in 1 to 16 colors or various shadings. The screen is divided in a matrix array. Each

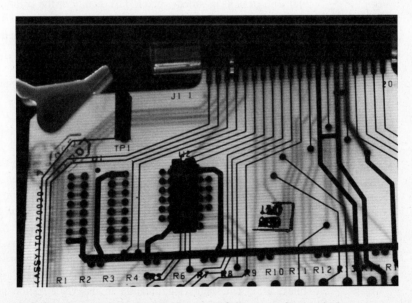

A design element circuit board. (Courtesy of Grumman Corporation)

Engineers still need to create and to read blueprints. The combination of visual and written abilities is very important in communicating with other professionals. (Courtesy of Xerox)

element, or cell, is called a pixel. The quality of the color display is determined by the number of pixels the computer can control. Some CRTs have a matrix of 4096 by 4096 pixels, but most have fewer elements; hence the resolution will not be as clear.

Using CAD systems for drawing requires a change in your orientation. When you draw a line between two points, you manually start at point x and draw the line to point y. Using a computer you define the points, x and y, and direct the computer to connect them with a straight, elliptical, or hyperbolic line. In addition the designer can assign line intensity, width, and coloration as well as other attributes.

Probably the CAD system you will first encounter in college is one using a wire-frame model. This is the simplest, most fundamental display of a three-dimensional model, where each edge appears as a line and the surfaces are transparent. Lines that are normally hidden are visible and must be removed by the designer. Some programs have hidden-line removal commands that create a quite realistic model. Figure 13.13 shows a three-dimensional model with and without hidden lines removed. Notice that the rendition (a), with all the lines included, is very difficult to visualize three-dimensionally. Removing the hidden lines, as in (b), helps significantly in the visualization process. While these

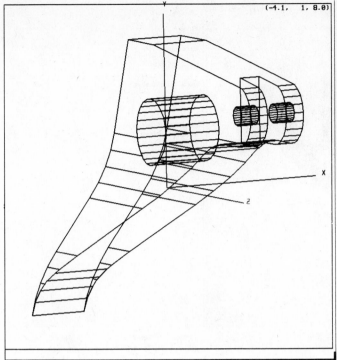

(-4.1,   1, 8.0)

DRAWING TYPE MENU

Return to
Display Menu

Wireframe

Haloed Line

Back Face Removal

Hidden Line

Hidden Surface
Constant Shading
Hidden Surface
Gouraud Shading

2D Picture Options

Lighting Options

FIGURE 13.13 (a) A wire-frame drawing of a machine part.

Drawing
Control
Help   Return to
            Displ Menu

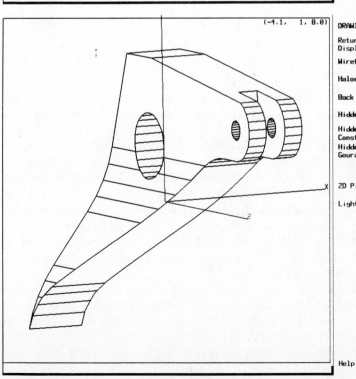

(-4.1,   1, 8.0)

DRAWING TYPE MENU

Return to
Display Menu

Wireframe

Haloed Line

Back Face Removal

Hidden Line

Hidden Surface
Constant Shading
Hidden Surface
Gouraud Shading

2D Picture Options

Lighting Options

FIGURE 13.13 (b) A wire-frame drawing of the same machine part, with the hidden lines removed. Notice the clarity you gain.

Drawing
Control
Help   Return to
            Displ Menu

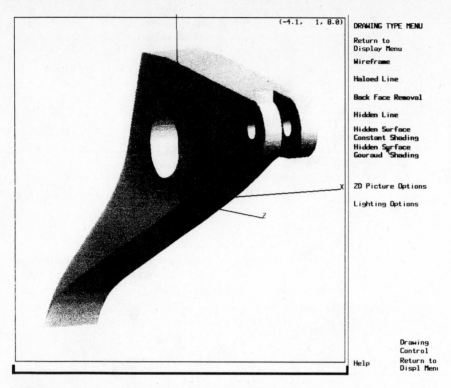

(-4.1, 1, 8.0)

DRAWING TYPE MENU

Return to
Display Menu

Wireframe

Haloed Line

Back Face Removal

Hidden Line

Hidden Surface
Constant Shading
Hidden Surface
Gouraud Shading

2D Picture Options

Lighting Options

Help

Drawing
Control
Return to
Displ Menu

FIGURE 13.13 (c) Solid modeling showing surfaces and shading of the same machine part.

models are very useful in drafting work and other two-dimensional applications, there are limitations in not having the third dimension. The computer cannot distinguish between inside and outside areas, and surface information is missing. Both are important to the manufacturing of three-dimensional objects.

The next level of CAD/CAM allows the surfaces to be modeled for certain objects. This technique works particularly well with thin materials, such as sheet metal, where the difference between the inner and outer surfaces is not significant. The surface model is generated by creating a wire-frame model and placing planes between the wire-frame edges. The techniques are sufficiently sophisticated to allow complicated curved surfaces, including fillets, to be modeled. Figure 13.13c illustrates the machine part with surfaces shown, the clearest rendition for us.

Solid modeling is the most sophisticated of graphics display; the part can be modeled with internal and external surfaces and any intermediate parts, all with different colorations. The quality of picture is excellent, with all surface nuances shown. These sophisticated workstations can light the object from different angles, producing very artistic effects. It is truly a blending of art and engineering.

Not only can the object be created, it can be analyzed using a finite element program. The finite element method is a numerical method of solving a variety of physical problems, from heat transfer to stress analysis. For most applications an exact mathematical solution to the equations modeling the problem is impossible, and approximate techniques, such as the finite element method, are employed. In this situation the total region, such as an airfoil (a thin plate) attached to a frame on an airplane, as shown in Figure 13.14a, is divided into small elements as in Figure 13.14b. The equations modeling the various physical parameters are written for each element, in simplified form because a small section is considered, and solved with the constraint that the solutions be consistent between the adjacent small elements. You can imagine that many equations must be solved simultaneously, requiring the processing power now available in computers.

The engineer can vary the element materials, their thickness, and method of joining in some cases, and can determine the stress at critical points. In this way the design can be modified as needed. In creating the design initially, the engineer drew the object with the aid of a CAD system. The computer contains this information in terms of coordinates defining the object. This information is transferred to the finite element program, so the engineer does not have to key in the coordinates again. The element size is dictated by the complexity of the shape where there are rapidly changing gradients in force or temperature. In these regions the grid size must be smaller than in regions of slowly changing gradients. Obviously, the smaller the elements the greater the required computing time and the greater the number of equations and boundary conditions that must be matched.

The F14-A Plus can fly at over twice the speed of sound (Mach 2.0+) and is the product of some of the most sophisticated engineering design and analysis in the world. (Courtesy of Grumman Corporation)

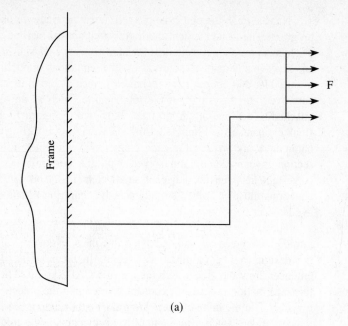

(a)

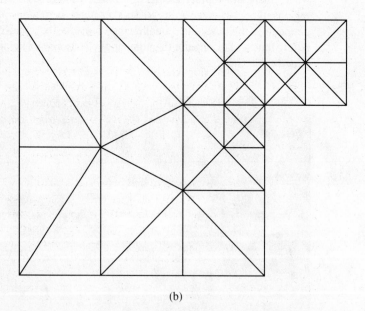

(b)

FIGURE 13.14 (a) A thin uniform plate, attached to a frame and with a uniform force acting on it. (b) A finite element grid pattern for the plate. The element size varies as a function of the change in stress gradient in the plate, with a smaller grid size in areas of high stress gradients.

We have now reached the CAM part of CAD/CAM. The design is completed and needs to be manufactured. The information about the design component, its CAD data file, can be sent to the computer controlling the manufacturing process, often a manufacturing robot. This is currently atypical, because the CAD to CAM interface is still cumbersome. The data from the CAD system must be compatible with the CAM system, and the manufacturer needs to invest in the robots and automated manufacturing systems that use the CAD in the first place. Not only must the company have the financial ability to fund such a conversion, but it requires a change, and decrease, in the work force. There are obvious social and political ramifications to such changes.

Data Bases and Data-Base Processing

Information is key to making engineering and management decisions. Knowledge about new products, sales, manufacturing costs, inventory, parts, and personnel becomes important. Not only are these subjects individually important, but their interrelationships can be extremely valuable at times. Knowing the in-house inventory of parts may reduce the time required to undertake a new product line. Data-base technology allows associated data to be processed as a whole.

Rather than considering an industrial plant, let's consider an example involving your current environment, a university. Different areas of the university need related specific information each semester. For instance, information about the faculty teaching load is required to generate their paychecks, while some of this information is needed to schedule who is teaching what, when, and where. Related to the class schedules are the student data regarding who is attending what classes, so grades may be given. Figure 13.15 illustrates these three application files and the data that are required in them. Each file contains its own data, even though the same data will appear in more than one file.

Data-base problems occur when you want information that crosses file boundaries, such as the average salary of faculty teaching a certain course. This requires information from the faculty file and the class data file. There is no certainty that the information in one file will be formatted in such a way as to be accessible by another file. Sometimes it is simply not worth the effort to get this information from the computer, unless the information is stored in a data base (Figure 13.16).

In this situation a data-base management system (DBMS) acts as system librarian. It stores the data and its description and retrieves this data for the applications programs as needed. The applications programs have not changed their function; they are just integrated into a whole by the DBMS. Several advantages spring to mind regarding DBMS: you can obtain more information from a given piece of data, as it may be correlated in a variety of ways; data duplication is reduced or eliminated; and inconsistencies in the data are eliminated, such as student or faculty name changes. The DBMS is more

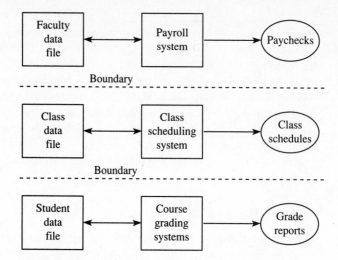

FIGURE 13.15 Three application files that are used in universities. There is no communication among the files.

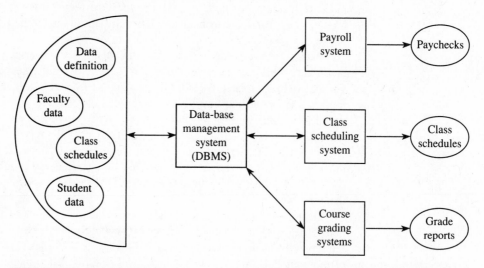

FIGURE 13.16 An integrated data-base management system allows information sharing. The boundaries between the data files are eliminated.

complex than a file system, and the applications programs need to be more sophisticated to handle the "generic" data. Because DBMS is an integrated system, failure in one part of the system can cause the entire system to be down, and in case of failure, recovery is more difficult. Backing up files is essential. The primary disadvantage to adopting DBMS is the conversion of data to the data-base system format and revising specific applications programs to accept the data in a more sophisticated form. However, the advantages of greater information gain are driving industry to adopt DBMS more fully, particularly as it can now be run on mainframes and minicomputers.

13.8 COMPUTER PROGRAMMING LANGUAGES AND FLOWCHARTING

When digital computers were first developed in the 1940s, and for about the following ten years, people had to communicate with the computer in machine language. Whereas this language is handy for computers, it is decidedly not easy to program and hence is error prone. Every computer manufacturer had its own machine language, and the computer instructions and memory addresses were specified in digits. Adding two values together might include these instructions:

150 58 410034

151 53 410038

152 50 421050

Machine languages, known as first-generation programming languages, were too difficult to use, and engineers soon developed a more symbolic language, assembly language programming, known as the second generation of languages. Each assembly language instruction is converted into machine language through the use of an assembler. Thus to add two values in assembly language the program might include the following instructions:

```
LDA X    LOAD X IN ACCUMULATOR
ADD Y    ADD Y TO ACCUMULATOR
STA Z    STORE ACCUM SUM AT Z
```

In this example X is loaded in the accumulator, Y is added to this value, and their sum, Z, is stored in a memory cell. This represents a tremendous advance over machine language programming, but it is still tedious. Enter the third-generation languages in the late 1950s and early 1960s, including BASIC, FORTRAN, and PL/1. In BASIC the summation problem is simply

100 Z = X + Y.

In the 1980s fourth-generation languages were developed, and they will continue to be created; they will be easier to use and more user-friendly, and will require less computer time to run. Table 13.5 shows the historical development of computer languages as well as their primary areas of application. The names of the languages are self-explanatory except for Ada, named for Lady Ada Lovelace, considered to be the world's first computer language developer in 1842. The computer was not electronic or electric, of course, but mechanical and known as Charles Babbage's analytic engine. The analytic engine could mechanically store 1000 numbers of 50 digits using ten position wheels. It was devised to store data and calculational results, perform mathematical operations, and input and output data. Lady Lovelace, the daughter of Lord Byron, devised a sequence of instructions for the engine to perform a set of complex calculations, the first computer program.

TABLE 13.5 PRIMARY HIGH-LEVEL PROGRAMMING LANGUAGES

Language name and meaning	Year introduced	Primary application areas
FORTRAN (*FOR*mula *TRAN*slation)	1957	Science and engineering
COBOL (*CO*mmon *B*usiness-*O*riented *L*anguage)	1959	Business data processing
ALGOL (*ALGO*rithmic *L*anguage)	1960	Science and engineering
APL (*A P*rogramming *L*anguage)	1962	Science and engineering (particularly time-sharing systems)
PL/1 (*P*rogramming *L*anguage/1)	1964	Business, science, and engineering
BASIC (*B*eginners *A*ll-purpose *S*ymbolic *I*nstruction *C*ode)	1965	Business, science, and engineering (particularly time-sharing systems)
Pascal (named after Blaise Pascal)	1971	Business, science, engineering, and education
C (the language after B)	1972	Scientific, engineering, and system software development
Ada (named after Augusta Ada Lovelace)	1980	Real-time, embedded computer systems

BASIC stands for beginner's all-purpose symbolic instruction code; it was developed at Dartmouth College in the 1960s. The word *beginner* is misleading, as it is a quite powerful language, very amenable to time-sharing systems and more importantly to personal computers. There are some variations to BASIC, as not all computers use the same version, but the versions are essentially the same. BASIC is very user-friendly and can handle sophisticated circuit analysis and mechanical analysis problems.

FORTRAN, Pascal, and BASIC form the core of current general-purpose engineering and scientific languages. FORTRAN is an acronym for "formula translation" and was developed with engineers and scientists in mind. During the time of its development keypunched cards were used, and its structure coordinates with this method of inputting data and formulae. Many FORTRAN programs are used in industry on mainframe computers, so the move from it to other languages will be slow. However, much of the program development for personal computers uses other languages, often BASIC, so a countervailing force is being created in the marketplace. This is not a great obstacle for you; when you learn one of these languages, switching to another is not difficult. The logic sequencing is the same.

Flowcharting

Flowcharting helps in developing this logic sequencing. A number of symbols are used, but here we consider only four symbols, as illustrated in Figure 13.17. The advantages to flowcharting include (1) displaying the logical steps in the problem solution, (2) allowing compact communication of algorithms and/or solutions to others, and (3) permanently recording the problem solution for later reference. The greatest drawback to flowcharting is that no one likes to do it. Usually the charts are done after the program is written and are no use in planning the program, their fundamental purpose.

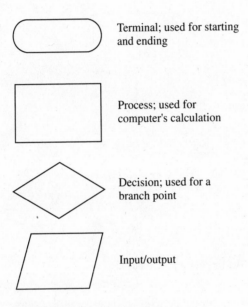

Terminal; used for starting and ending

Process; used for computer's calculation

Decision; used for a branch point

Input/output

FIGURE 13.17 Flowchart symbols.

EXAMPLE
13.6

Create a flowchart depicting the simple interest I annually from the equation $I = Prt$, where P is the principal, r is the interest rate, and t is the time in years for a period of ten years.

SOLUTION: Figure 13.18 flowcharts this equation.

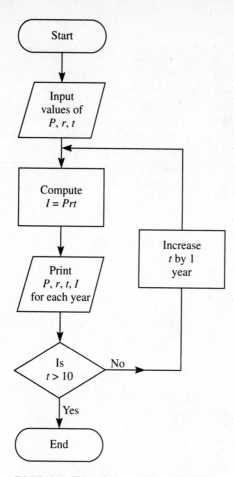

FIGURE 13.18 The solution to Example 13.6.

REFERENCES

1. Pollard, L.H. *Computer Design and Architecture*. Prentice-Hall, Englewood Cliffs, NJ, 1990.
2. Roth, C.H., Jr. *Fundamentals of Logic Design*. 3rd ed. West Publishing, St. Paul, MN, 1985.
3. Terrell, D.L. *Microprocessor Technology*. Reston Publishing, Reston, VA, 1983.

PROBLEMS

13.1 Write the following decimal numbers in powers of ten notation.
 a) 1990   **b)** 22.22   **c)** 0.1911

13.2 Write the following binary numbers in powers of two notation.
 a) 11001   **b)** 100100111   **c)** 0.11101

13.3 Convert the following decimal numbers to their binary equivalent.
a) 296   **b)** 16 941   **c)** 3.1416   **d)** 0.1110

13.4 Convert the following binary numbers to their decimal equivalent.
a) 101.101   **b)** 10001001   **c)** 0.1110

13.5 Convert the following decimal numbers to their octal and hexadecimal equivalent.
a) 19   **b)** 2015   **c)** 92.20   **d)** 0.824

13.6 Convert the following binary numbers into hexadecimal equivalents.
a) 1010101101   **b)** 1011.01   **c)** 0.101101

For problems 13.7, 13.8 and 13.9 check your results by converting to the decimal system.

13.7 Add the following pairs of binary numbers.
a) 1001   **b)** 101   **c)** 1101   **d)** 10010010
    101   111   10   10001110

13.8 Subtract the following pairs of binary numbers.
a) 111   **b)** 1001   **c)** 1111   **d)** 1100011
    11   111   1001   1011100

13.9 Multiply the following pairs of binary numbers.
a) 1101   **b)** 10000   **c)** 110   **d)** 110011
    100   10110   11   100101

13.10 Prepare a truth table for each of the circuits shown.

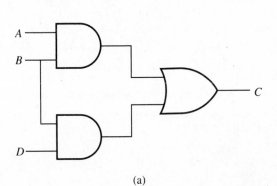

(a)

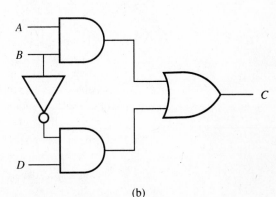

(b)

13.11 You have been assigned the task of creating the logic circuit for a bank alarm. The alarm is to sound, logic 1 state, if the master switch is on, if the bank door is open, and if the safe door is open.

13.12 Design the logic circuit for starting an emergency diesel generator. For the generator to start, the generator must be disconnected from the bus, there must be sufficient starting air pressure, and the fuel tank must indicate that there is oil. When these conditions are met, then the diesel may be started. Presume that there are sensors to determine these values.

13.13 The figure illustrates a schematic diagram for a half adder; it will add any two binary digits and produce the resulting summation as the outputs. Show that it will correctly add all combinations of binary input.

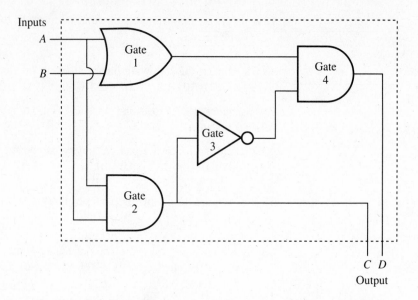

13.14 Assuming that a spreadsheet program is available, create a problem to solve the following: You are in charge of billing various items the engineering school made for sale during homecoming. There are five items, A, B, C, D, and E, at prices of $20, $15, $10, $5, and $1 each. If any purchaser buys more than five items of a given kind, the purchaser receives a 15 percent discount. A sales tax of 6 percent must be collected, and the bill should show the number of items purchased. Write a spreadsheet program for this purpose.

13.15 Using a spreadsheet, solve the equation $z^3x^2 (1 + 2y^2)$ for $x = 3.0$, $z = -3.5$, and $y = 5.1$.

13.16 Using a spreadsheet, solve the set of equations.

$$2x + y = 7$$
$$3x + y = 9$$

13.17 Create a flowchart to calculate the grades for a class of students where the student ID and four test grades are entered and the student and class final averages are computed.

13.18 Create a flowchart for the solution of the hypotenuse of a right triangle for various combinations of sides of the triangle.

NATIONAL SOCIETY OF PROFESSIONAL ENGINEERS CODE OF ETHICS FOR ENGINEERS

Preamble

Engineering is an important and learned profession. The members of the profession recognize that their work has a direct and vital impact on the quality of life for all people. Accordingly, the services provided by engineers require honesty, impartiality, fairness and equity, and must be dedicated to the protection of the public health, safety and welfare. In the practice of their profession, engineers must perform under a standard of professional behavior which requires adherence to the highest principles of ethical conduct on behalf of the public, clients, employers and the profession.

Fundamental Canons

Engineers, in the fulfillment of their professional duties, shall:

1. Hold paramount the safety, health and welfare of the public in the performance of their professional duties.
2. Perform services only in areas of their competence.
3. Issue public statements only in an objective and truthful manner.
4. Act in professional matters for each employer or client as faithful agents or trustees.
5. Avoid improper solicitation of professional employment.

Rules of Practice

1. Engineers shall hold paramount the safety, health and welfare of the public in the performance of their professional duties.

   (a) Engineers shall at all times recognize that their primary obligation is to protect the safety, health, property and welfare of the public. If their professional judgment is overruled under circumstances where the safety, health, property or welfare of the public are endangered, they shall notify their employer or client and such other authority as may be appropriate.

   (b) Engineers shall approve only those engineering documents which are safe for public health, property and welfare in conformity with accepted standards.

   (c) Engineers shall not reveal facts, data or information obtained in a professional capacity without the prior consent of the client or employer except as authorized or required by law or this Code.

   (d) Engineers shall not permit the use of their name or firm name nor associate in business ventures with any person or firm which they have reason to believe is engaging in fraudulent or dishonest business or professional practices.

   (e) Engineers having knowledge of alleged violations of this Code shall cooperate with the proper authorities in furnishing such information or assistance as may be required.

2. Engineers shall perform services only in the areas of their competence.

   (a) Engineers shall undertake assignments only when qualified by education or experience in the specific technical field involved.

   (b) Engineers shall not affix their signatures to any plans or documents dealing with subject matter in which they lack competence, nor to any plan or document not prepared under their direction and control.

   (c) Engineers may accept an assignment outside of their fields of competence to the extent that their services are restricted to those phases of the project in which they are qualified, and to the extent that they are satisfied that all other phases of such project will be performed by registered or otherwise qualified associates, consultants, or employees, in which case they may then sign the documents for the total project.

3. Engineers shall issue public statements only in an objective and truthful manner.

   (a) Engineers shall be objective and truthful in professional reports, statements or testimony. They shall include all relevant and pertinent information in such reports, statements or testimony.

   (b) Engineers may express publicly a professional opinion on technical subjects only when the opinion is founded upon adequate knowledge of the facts and competence in the subject matter.

   (c) Engineers shall issue no statements, criticisms or arguments on technical matters which are inspired or paid for by interested parties, unless they have prefaced their comments by explicitly identifying the interested parties on whose behalf they are speaking, and by revealing the existence of any interest the engineers may have in the matters.

4. Engineers shall act in professional matters for each employer or client as faithful agents or trustees.

(a) Engineers shall disclose all known or potential conflicts of interest to their employers or clients by promptly informing them of any business association, interest or other circumstances which could influence or appear to influence their judgment or the quality of their services.

(b) Engineers shall not accept compensation, financial or otherwise, from more than one party for services on the same project, or for services pertaining to the same project, unless the circumstances are fully disclosed to, and agreed to, by all interested parties.

(c) Engineers shall not solicit or accept financial or other valuable consideration, directly or indirectly, from contractors, their agents, or other parties in connection with work for employers or clients for which they are responsible.

(d) Engineers in public service as members, advisors or employees of a governmental body or department shall not participate in decisions with respect to professional services solicited or provided by them or their organizations in private or public engineering practice.

(e) Engineers shall not solicit or accept a professional contract from a governmental body on which a principal or officer of their organization serves as a member.

5. Engineers shall avoid improper solicitation of professional employment.

(a) Engineers shall not falsify or permit misrepresentation of their, or their associates', academic or professional qualifications. They shall not misrepresent or exaggerate their degree of responsibility in or for the subject matter of prior assignments. Brochures or other presentations incident to the solicitation of employment shall not misrepresent pertinent facts concerning employers, employees, associates, joint ventures or past accomplishments with the intent and purpose of enhancing their qualifications and their work.

(b) Engineers shall not offer, give, solicit or receive, either directly or indirectly, any political contribution in an amount intended to influence the award of a contract by public authority, or which may be reasonably construed by the public of having the effect or intent to influence the award of a contract. They shall not offer any gift, or other valuable consideration, in order to secure work. They shall not pay a commission, percentage or brokerage fee in order to secure work except to a bona fide employee or bona fide established commercial or marketing agencies retained by them.

Professional Obligations

1. Engineers shall be guided in all their professional relations by the highest standards of integrity.

   (a) Engineers shall admit and accept their own errors when proven wrong and refrain from distorting or altering the facts in an attempt to justify their decisions.

   (b) Engineers shall advise their clients or employers when they believe a project will not be successful.

   (c) Engineers shall not accept outside employment to the detriment of their regular work or interest. Before accepting any outside employment they will notify their employers.

   (d) Engineers shall not attempt to attract an engineer from another employer by false or misleading pretenses.

   (e) Engineers shall not actively participate in strikes, picket lines, or other collective coercive action.

   (f) Engineers shall avoid any act tending to promote their own interest at the expense of the dignity and integrity of the profession.

2. Engineers shall at all times strive to serve the public interest.

   (a) Engineers shall seek opportunities to be of constructive service in civic affairs and work for the advancement of the safety, health and well-being of their community.

   (b) Engineers shall not complete, sign or seal plans and/or specifications that are not of a design safe to the public health and welfare and in conformity with accepted engineering standards. If the client or employer insists on such unprofessional conduct, they shall notify the proper authorities and withdraw from further services on the project.

   (c) Engineers shall endeavor to extend public knowledge and appreciation of engineering and its achievements and to protect the engineering profession from misrepresentation and misunderstanding.

3. Engineers shall avoid all conduct or practice which is likely to discredit the profession or deceive the public.

   (a) Engineers shall avoid the use of statements containing a material misrepresentation of fact or omitting a material fact necessary to keep statements from being misleading; statements intended or likely to create an unjustified expectation; statements containing prediction of future success; statements containing an opinion as to the quality of the engineers' services; or statements intended or likely to attract clients by the use of showmanship, puffery, or self-laudation, including the use of slogans, jingles, or sensational language or format.

   (b) Consistent with the foregoing; engineers may advertise for recruitment of personnel.

(c) Consistent with the foregoing; engineers may prepare articles for the lay or technical press, but such articles shall not imply credit to the author for work performed by others.

4. Engineers shall not disclose confidential information concerning the business affairs or technical processes of any present or former client or employer without his consent.

  (a) Engineers in the employ of others shall not without the consent of all interested parties enter promotional efforts or negotiations for work or make arrangements for other employment as a principal or to practice in connection with a specific project for which the engineer has gained particular and specialized knowledge.

  (b) Engineers shall not, without the consent of all interested parties, participate in or represent an adversary interest in connection with a specific project or proceeding in which the engineer has gained particular specialized knowledge on behalf of a former client or employer.

5. Engineers shall not be influenced in their professional duties by conflicting interests.

  (a) Engineers shall not accept financial or other considerations, including free engineering designs, from material or equipment suppliers for specifying their product.

  (b) Engineers shall not accept commissions or allowances, directly or indirectly, from contractors or other parties dealing with clients or employers of the engineer in connection with work for which the engineer is responsible.

6. Engineers shall uphold the principle of appropriate and adequate compensation for those engaged in engineering work.

  (a) Engineers shall not accept remuneration from either an employee or employment agency for giving employment.

  (b) Engineers, when employing other engineers, shall offer a salary according to professional qualifications and the recognized standards in the particular geographical area.

7. Engineers shall not compete unfairly with other engineers by attempting to obtain employment or advancement or professional engagements by taking advantage of a salaried position, by criticizing other engineers, or by other improper or questionable methods.

  (a) Engineers shall not request, propose, or accept a professional commission on a contingent basis under circumstances in which their professional judgment may be compromised.

  (b) Engineers in salaried positions shall accept part-time engineering work only at salaries not less than that recognized as standard in the area.

  (c) Engineers shall not use equipment, supplies, laboratory or office facilities of an employer to carry on outside private practice without consent.

8. Engineers shall not attempt to injure, maliciously or falsely, directly or indirectly, the professional reputation, prospects, practice or employment of other engineers, nor indiscriminately criticize other engineers' work. Engineers who believe others are guilty of unethical or illegal practice shall present such information to the proper authority for action.

(a) Engineers in private practice shall not review the work of another engineer for the same client, except with the knowledge of such engineer, or unless the connection of such engineer with the work has been terminated.

(b) Engineers in governmental, industrial or educational employ are entitled to review and evaluate the work of other engineers when so required by their employment duties.

(c) Engineers in sales or industrial employ are entitled to make engineering comparisons of represented products with products of other suppliers.

9. Engineers shall accept personal responsibility for all professional activities.

(a) Engineers shall conform with state registration laws in the practice of engineering.

(b) Engineers shall not use association with a nonengineer, a corporation or a partnership, as a "cloak" for unethical acts, but must accept personal responsibility for all professional acts.

10. Engineers shall give credit for engineering work to those to whom credit is due, and will recognize the proprietary interests of others.

(a) Engineers shall, whenever possible, name the person or persons who may be individually responsible for designs, inventions, writings or other accomplishments.

(b) Engineers using designs supplied by a client recognize that the designs remain the property of the client and may not be duplicated by the engineer for others without express permission.

(c) Engineers, before undertaking work for others in connection with which the engineer may make improvements, plans, designs, inventions, or other records which may justify copyrights or patents, should enter into a positive agreement regarding ownership.

(d) Engineers' designs, data, records, and notes referring exclusively to an employer's work are the employer's property.

11. Engineers shall cooperate in extending the effectiveness of the profession by interchanging information and experience with other engineers and students, and will endeavor to provide opportunity for professional development and advancement of engineers under their supervision.

(a) Engineers shall encourage engineering employees' efforts to improve their education.

(b) Engineers shall encourage engineering employees to attend and present papers at professional and technical society meetings.

(c) Engineers shall urge engineering employees to become registered at the earliest possible date.

(d) Engineers shall assign a professional engineer duties of a nature to utilize full training and experience, insofar as possible, and delegate lesser functions to subprofessionals or to technicians.

(e) Engineers shall provide a prospective engineering employee with complete information on working conditions and proposed status of employment, and after employment will keep employees informed of any changes.

As Revised March 1985

"By order of the United States District Court for the District of Columbia, former Section 11(c) of the NSPE Code of Ethics prohibiting competitive bidding, and all policy statements, opinions, rulings or other guidelines interpreting its scope, have been rescinded as unlawfully interfering with the legal right of engineers, protected under the antitrust laws, to provide price information to prospective clients; accordingly, nothing contained in the NSPE Code of Ethics, policy statements, opinions, rulings or other guidelines, prohibits the submission of price quotations or competitive bids for engineering services at any time or in any amount."

Statement by NSPE Executive Committee

In order to correct misunderstandings which have been indicated in some instances since the issuance of the Supreme Court decision and the entry of the final judgment, it is noted that in its decision of April 25, 1978, the Supreme Court of the United States declared: "The Sherman Act does not require competitive bidding."

It is further noted that as made clear in the Supreme Court decision:

1. Engineers and firms may individually refuse to bid for engineering services.
2. Clients are not required to seek bids for engineering services.
3. Federal, state and local laws governing procedures to procure engineering services are not affected, and remain in full force and effect.
4. State societies and local chapters are free to actively and aggressively seek legislation for professional selection and negotiation procedures by public agencies.
5. State registration board rules of professional conduct, including rules prohibiting competitive bidding for engineering services, are not affected and remain in full force and effect. State registration boards with

authority to adopt rules of professional conduct may adopt rules governing procedures to obtain engineering services.

6. As noted by the Supreme Court, "nothing in the judgment prevents NSPE and its members from attempting to influence governmental action. . . ."

Note: In regard to the question of application of the Code to corporations vis-à-vis real persons, business form or type should not negate nor influence conformance of individuals to the Code. The Code deals with professional services, which services must be performed by real persons. Real persons in turn establish and implement policies within business structures. The Code is clearly written to apply to the engineer and it is incumbent on a member of NSPE to endeavor to live up to its provisions. This applies to all pertinent sections of the Code.

SOURCE: NSPE Publication No. 1102 as revised January 1987

CONVERSION FACTORS

Multiply	By	To Obtain
abamperes	10	amperes
abcoulombs	10	coulombs
abfarads	10^9	farads
abhenries	10^{-9}	henries
abohms	10^{-9}	ohms
abvolts	10^{-8}	volts
acres	43 560	square feet
"	4047	square meters
"	4840	square yards
acre-feet	43 560	cubic feet
" "	3.259×10^5	gallons (U.S. liquid)
amperes	1	coulombs per second
angstroms	10^{-10}	meters
ares	0.02471	acres
"	100	square meters
astronomical units	1.496×10^{11}	meters
atmospheres	76	centimeters of mercury
"	29.92	inches of mercury
"	33.90	feet of water
"	14.70	pounds per square inch (force)
"	1.013×10^5	pascals
bars	0.9872	atmospheres
"	10^6	dynes per square centimeter
"	14.51	pounds per square inch (force)
"	10^5	pascals

Multiply	By	To Obtain
barrels (petroleum)	42	gallons (U.S. liquid)
becquerels (radioactivity)	1	disintegrations per second
British thermal units	778.2	foot-pounds force
" " "	1055	joules
" " "	2.931×10^{-4}	kilowatt hours
Btu per minute	17.58	watts
Btu per pound	2326	joules per kilogram
bushels	1.244	cubic feet
"	0.035 24	cubic meters
"	4	pecks
calories	4.187	joules
"	10^{-3}	kilocalories
calories per gram	4187	joules per kilogram
Calories (kilocalories)	4187	joules
candelas per square meter	3.141×10^{-4}	lamberts
carats (metric)	2×10^{-4}	kilograms (mass)
centimeters	0.3937	inches
centimeters of mercury	0.013 16	atmospheres
" " "	0.4461	feet of water
" " "	0.1935	pounds per square inch (force)
" " "	1333	pascals
circular mils	5.067×10^{-6}	square centimeters
" "	7.854×10^{-7}	square inches
cords	8 ft. $\times$ 4 ft. $\times$ 4 ft.	cubic feet
coulombs (quantity of electricity)	1	ampere-seconds
cubic centimeters	3.531×10^{-5}	cubic feet
" "	6.102×10^{-2}	cubic inches
" "	10^{-6}	cubic meters
" "	10^{-3}	liters
" "	2.642×10^{-4}	gallons (U.S. liquid)
cubic feet	2.832×10^{4}	cubic centimeters
" "	1728	cubic inches
" "	0.028 32	cubic meters
" "	0.037 04	cubic yards
" "	7.481	gallons (U.S. liquid)
" "	28.32	liters
cubic inches	16.39	cubic centimeters
" "	5.787×10^{-4}	cubic feet
" "	1.639×10^{-5}	cubic meters
" "	2.143×10^{-5}	cubic yards
" "	4.329×10^{-3}	gallons (U.S. liquid)
cubic meters	35.31	cubic feet
" "	61 024	cubic inches
" "	1.308	cubic yards
" "	264.2	gallons (U.S. liquid)

Multiply	By	To Obtain
cubic yards	27	cubic feet
" "	46 656	cubic inches
" "	0.7646	cubic meters
" "	202.0	gallons (U.S. liquid)
curies	3.7×10^{10}	becquerels
days	24	hours
"	1440	minutes (time)
"	8.640×10^4	seconds (time)
degrees (angle)	60	minutes (angle)
" "	0.017 45	radians
degrees Fahrenheit	—	degrees Celsius: $t_c = (t_F - 32)/1.8$
degrees Celsius	—	kelvin: $T = t_c + 273.15$ K
degrees Fahrenheit	—	kelvin: $T = t_F + 459.67°R)/1.8$
degrees Rankine	—	kelvin: $T = T_R/1.8$
degrees per second (angle)	0.1667	revolutions per minute
degrees Kelvin (see kelvin)		
density: pounds-mass/in.^{3}	27 680	kilograms per cubic meter (mass)
drams	1.772	grams force
"	0.0625	ounces force
dynes	1.020×10^{-3}	grams force
"	7.233×10^{-5}	poundals
"	2.248×10^{-6}	pounds force
"	1	gram-centimeters/s^2 (mass)
"	10^{-5}	newtons
electron volts	1.602×10^{-19}	joules
ergs	9.479×10^{-11}	British thermal units
"	7.378×10^{-8}	foot-pounds force
"	10^{-7}	joules
"	1	dyne-centimeters
ergs per second	1.341×10^{-10}	horsepower
" " "	10^{-7}	watts
farads (electric capacitance)	1	coulombs per volt
fathoms	6	feet
feet	0.3048	meters
feet per second	0.3048	meters per second
feet of water	0.029 50	atmospheres
" " "	0.8827	inches of mercury
" " "	0.4336	pounds per square inch (force)
feet of water (39.2°F)	2989	pascals
foot-candles	10.76	lumens per square meter (lux)
" "	10.76	lux
" "	1	lumens per square foot

Multiply	By	To Obtain
foot-pounds force	1.285×10^{-3}	British thermal units
" " "	1.356×10^{7}	ergs
" " "	1.356	joules
force: lbf	4.448	newtons
force: kgf ("kilopond")	9.807	newtons
force (1 kg·m/s²)	1	newtons
frequency (1/s)	1	hertz
furlongs	40	rods
gallons (U.S. liquid)	3.785×10^{-3}	cubic meters
" " " "	0.1337	cubic feet
" " " "	231	cubic inches
" " " "	4	quarts (U.S. liquid)
gallons (U.S. dry)	4.405×10^{-3}	cubic meters
gallons (U.K. liquid)	4.546×10^{-3}	cubic meters
gals (unit of acceleration)	10^{-2}	meters per second per second
gammas (mass)	10^{-9}	kilograms (mass)
gammas (magnetic flux density)	10^{-9}	teslas
gausses	10^{-4}	teslas
gills	0.25	pints (U.S. liquid)
grads	1.571×10^{-2}	radians
grains	1.429×10^{-4}	pounds
grams	10^{-3}	kilograms
grams force	0.035 27	ounces force
" "	0.032 15	ounces force (troy)
" "	2.204×10^{-3}	pounds force
hectares	2.471	acres
"	10^{4}	square meters
henries (inductance)	1	webers per ampere
horsepower	42.41	British thermal units per minute
" "	33 000	foot-pounds per minute (force)
" "	550	foot-pounds per second (force)
" "	745.7	watts
horsepower-hour	2.684×10^{6}	joules
inches	2.540	centimeters
"	2.540×10^{-2}	meters
inches of mercury (32°F)	0.033 42	atmospheres
" " " "	0.4912	pounds per square inch (force)
" " " "	3.386×10^{3}	pascals
inches of mercury (60°F)	3.377×10^{3}	pascals
joules (energy, work, heat)	1	newton meters
"	9.478×10^{-4}	British thermal units
"	0.7376	foot-pounds force
"	2.778×10^{-4}	watt-hours
"	0.2388	calories

Multiply	By	To Obtain
joules	2.388×10^{-4}	kilocalories
joules per kilogram	4.300×10^{-4}	Btu per pound
kelvin	—	degrees Celsius:
		$t_C = T - 273.15$ K
"	—	degrees Fahrenheit:
		$t_F = 1.8\,T - 459.67$ R
"	—	degrees Rankine: $T_R = 1.8T$
kilocalories	4.187×10^3	joules
"	10^3	calories
kilograms force (kgf)	70.93	poundals
" " "	2.205	pounds force
" " "	9.807	newtons
kilograms mass (kg)	1	kilograms
" " "	0.068 54	slugs (mass)
" " "	2.205	pounds mass
kilograms per cubic meter	0.062 43	pounds per cubic foot
kilograms per square meter (force)	1.422×10^{-3}	pounds per square inch (force)
kilometers	3281	feet
"	0.6214	miles
"	10^3	meters
kiloponds (kgf)	9.807	newtons
kilowatts	10^3	watts
kilowatt-hours	3.600×10^6	joules
kips (1000 lbf)	4.448×10^3	newtons
kips per square inch	6.895×10^6	pascals
knots (international)	1.151	miles per hour
" "	0.5144	meters per second
lamberts	3183	candelas per square meter
leagues (nautical)	5556	meters
leagues (U.S. survey)	4828	meters
light years	9.461×10^{15}	meters
liters	10^{-3}	cubic meters
"	0.035 31	cubic feet
"	0.2642	gallons (U.S. liquid)
"	10^3	cubic centimeters
lumens (luminous flux)	1	candela-steradians
lumens per square foot	1	foot-candles
lumens per square meter	1	lux
lux (illuminance)	1	lumens per square meter
lux (lm/m^2)	0.0929	foot-candles
mass: lbm	0.4536	kilograms (mass)
maxwells	10^{-8}	webers
meters	1.094	yards
"	3.281	feet

Multiply	By	To Obtain
meters	39.37	inches
"	6.214×10^{-4}	miles (U.S. survey)
meters per second	3.281	feet per second
metric carats	2×10^{-4}	kilograms
metric tons (tonnes)	10^3	kilograms
mhos	1	siemens
microns	10^{-6}	meters
miles (nautical)	1852	meters
miles (U.S. survey)	1609	meters
" " "	5280	feet
" " "	1.609	kilometers
" " "	1760	yards
miles per hour	88	feet per minute
" " "	0.8688	knots (international)
milliamperes	10^{-3}	amperes
millibars	10^2	pascals
millimeters	0.039 37	inches
"	10^{-3}	meters
millimeters of mercury (0°C)	133.3	pascals
millivolts	10^{-3}	volts
mils	10^{-3}	inches
miner's inches	1.5	cubic feet per minute
minutes (angle)	2.909×10^{-4}	radians
nautical miles	1852	meters
newtons	1	kilogram-meters per second per second (kg·m/s²)
"	0.2248	pounds-force
"	10^5	dynes
"	0.1020	kilograms force
"	7.233	poundals
newton meters	0.7376	pound-feet (force)
oersteds	79.58	amperes per meter
ohms (electric resistance)	1	volts per ampere
ounces (troy)	0.083 33	pounds (troy)
" "	1.097	ounces (avoirdupois)
ounces force	0.2780	newtons
" "	28.35	grams force
" "	0.0625	pounds force
ounces force (troy)	31.10	grams force
parsecs	3.084×10^{16}	meters
pascals (pressure, stress)	1	newtons per square meter
"	0.9872×10^{-5}	atmospheres
"	2.953×10^{-4}	inches of mercury (32°F)
"	7.501×10^{-3}	millimeters of mercury (0°C) (torr)

Multiply	By	To Obtain
pascals	1.450×10^{-4}	pounds per square inch (force)
pecks (U.S.)	8.810×10^{-3}	cubic meters
pennyweights	1.555×10^{-3}	kilograms (mass)
picas (printer's)	4.218×10^{-3}	meters
pints (U.S. liquid)	4.732×10^{-4}	cubic meters
" (U.S. dry)	5.506×10^{-4}	cubic meters
points (printer's)	3.515×10^{-4}	meters
poises (absolute viscosity)	10^{-1}	pascal-seconds
poundals	0.1383	newtons
"	1.383×10^4	dynes
"	0.031 08	pounds force
pounds (avoirdupois)	7000	grains
pounds (troy)	0.8229	pounds (avoirdupois)
" "	5760	grains
pound-feet (force)	1.356	newton meters
pounds force (lbf)	453.7	grams force
" " "	16	ounces force
" " "	32.18	poundals
" " "	4.448	newtons
pounds mass (lbm)	0.4536	kilograms (mass)
pounds per cu ft	16.02	kilograms per cubic meter
pounds per sq in (psi)	0.068 03	atmospheres
" " " " "	2.036	inches of mercury
" " " " "	6895	pascals
" " " " "	6.895×10^{-3}	megapascals
pressure: psi	6895	pascals
pressure: atmospheres	1.013×10^5	pascals
quarts (U.S. liquid)	9.464×10^{-4}	cubic meters
" " "	0.2500	gallons (U.S. liquid)
radians	57.30	degrees (angle)
"	63.65	grads
"	0.1592	revolutions
rads (radiation dose absorbed)	10^{-2}	joules per kilogram (grays)
rods (U.S. survey)	16.5	feet
roentgens	2.580×10^{-4}	coulombs per kilogram
revolutions	2π	radians
sections (U.S. survey)	640	acres
"	2.590×10^6	square meters
siemens (electric conductance)	1	amperes per volt
"	1	mhos
slugs (mass)	14.59	kilograms (mass)
square centimeters	10^{-4}	square meters
statamperes	3.336×10^{-10}	amperes
statcoulombs	3.336×10^{-10}	coulombs
statfarads	1.113×10^{-12}	farads

Multiply	By	To Obtain
stathenries	8.988×10^{11}	henries
statohms	8.988×10^{11}	ohms
statvolts	299.8	volts
steres	1	cubic meters
stokes (kinematic viscosity)	10^{-4}	square meters per second
tablespoons	1.479×10^{-5}	cubic meters
teaspoons	4.929×10^{-6}	cubic meters
temperature (°C) + 273.15	1	absolute temperature (kelvin)
" " " + 17.78	1.8	temperature (°F)
temperature (°F) + 459.67	1	absolute temperature (°R)
" " " − 32	5/9	temperature (°C)
teslas (magnetic flux density)	1	webers per square meter
teslas	10^4	gausses
therms	10^5	British thermal units (Btu)
tonnes (metric tons)	10^3	kilograms
tons (long)	1016	kilograms
" "	2240	pounds
tons (metric)	10^3	kilograms
" "	2205	pounds
tons (short)	907.2	kilograms
" "	2000	pounds
tons (of refrigeration)	1.2×10^4	British thermal units per hour
tons (nuclear equivalent of TNT)	4.20×10^9	joules
torr (mm Hg, 0°C)	133.3	pascals
volts (electric potential)	1	watts per ampere
watts (power)	0.056 88	British thermal units per minute
" "	10^7	ergs per second
" "	1.341×10^{-3}	horsepower
" "	1	joules per second
watt-hours	3600	joules
webers (magnetic flux)	1	volt-seconds
"	10^8	maxwells
yards	0.9144	meters
"	3	feet

CALCULATORS

Hand-held calculators are invaluable for engineers and engineering students. For hundreds of years engineers have used some type of hand-held calculator. Slide rules are based on logarithms, which permit the mathematical operations of multiplication, division, and raising to powers. The oldest hand-held calculator, the abacus, is still used in Asia for quickly adding and subtracting. Electronic calculators can perform all of these operations and much more.

The calculator that you purchase as an engineering student will not be a lifelong investment. As you advance in your profession you will undoubtedly require a calculator with different, more, or fewer features. The general type of

A variety of hand-held calculators are available in algebraic and RPN form. (Courtesy of Hewlett-Packard)

calculator that you need as a student is a scientific calculator. The advances in electronics and semiconductors have permitted some hand-held calculators to be programmable in BASIC and to give graphic output. Some can even attach a small printer. We will look at features that are essential and note some of the advanced features that you may want. All of the scientific calculators come with a handbook that demonstrates how to use all the features of that particular model.

The following is a list of the features that many engineering faculty have found useful for use in undergraduate courses. (It assumes all have addition, subtraction, multiplication, and division operations.)

SIN	computes the sine of an angle; the inverse will compute the arcsine.
COS	computes the cosine of an angle; the inverse will compute the arccosine.
TAN	computes the tangent of an angle; the inverse will compute the arctangent.
LOG	computes the common logarithm (to the base 10) of a number; the inverse computes 10^x.
LN	computes the natural logarithm (to the base e) of a number; the inverse computes e^x.
y^x	computes the value of a number (y) raised to a positive, negative, or fractional power (x).
hyp	computes the hyperbolic functions: sinh, cosh, and tanh and their inverses.
x^2	computes the square of x.
$\sqrt{x}$	computes the square root of x.
$1/x$	computes the inverse of x.

Additional features that most scientific calculators have and that are useful include

Conversion between decimal, binary, octal, and hexadecimal number systems.

Statistical calculations, such as finding the standard deviation, mean, and median values.

Linear regression analysis.

Conversion from degrees to radians, from decimal notation to scientific notation.

Complex number addition, subtraction, multiplication, and division.

Most scientific calculators will have the aforementioned features. Before purchasing a calculator with more extensive features, ask yourself if you will need them in your course of study. In an educational setting computers are often available in laboratories or nearby for performing any large and/or repetitive calculations. The same is true for homework problems. In-class tests seldom, if ever, require a program to be run to determine a solution. The focus of test questions is towards concept understanding, demonstrated through the solution of fairly constrained problems due to time limitations. If you need to perform experiments in the field, however, and wish to check the accuracy of the measurements or the need for additional testing, a hand-held programmable calculator may prove very useful. Likewise, if computers are not readily available for situations you may typically encounter in labs and performing homework, a more powerful programmable calculator may be a wise purchase.

In addition to the calculator features, there are two general types of calculators, algebraic entry and RPN (reverse polish notation) entry. The difference may be illustrated by the problem of adding $3 + 2$. For algebraic-entry calculators we would push the 3 button, the addition sign, the 2 button, and the equal sign, and the number 5 would be displayed. In terms of registers, when 3 is pushed, the value of three enters the X register. Upon pushing the + button, this value moves to the Y register, leaving the X register empty for the next value. The number 2 fills the X register, and when you instruct the calculator what the next operation will be by pushing the equal sign, the resulting value of 5 is displayed in the X register. The Y register is now empty.

On calculators using RPN, most notably those manufactured by Hewlett-Packard, the calculator has an enter key, which enters the value into a register. The X register is considered the display register. When the enter key is pushed, the value from the display enters one of the other registers, typically denoted as Y, Z, and T registers. To add $3 + 2$, we push the 3 button, then the ENTER button. This enters 3 in the Y register. We push the 2 button, then ENTER. This enters 2 in the Z register. When we push the + button, the numbers in the two registers are added, with the output displayed as 5 in the X register.

Depending on the calculator you purchase, particular keying techniques must be used to assure that mathematical operations occur in the sequence you intend. The handbook accompanying the calculator will illustrate these in detail.

THE GREEK ALPHABET

A	α	Alpha
B	β	Beta
Γ	γ	Gamma
Δ	δ	Delta
E	ϵ	Epsilon
Z	ζ	Zeta
H	η	Eta
Θ	θ	Theta
I	ι	Iota
K	κ	Kappa
Λ	λ	Lambda
M	μ	Mu
N	ν	Nu
Ξ	ξ	Xi
O	o	Omicron
Π	π	Pi
P	ρ	Rho
Σ	σ	Sigma
T	τ	Tau
Υ	υ	Upsilon
Φ	ϕ	Phi
X	χ	Chi
Ψ	ψ	Psi
Ω	ω	Omega

USEFUL ALGEBRAIC FORMULAE

$$a^x a^y = a^{x+y}$$

$$\frac{a^x}{a^y} = a^{x-y} \text{ for } a \neq 0$$

$$(a^x)^y = a^{xy}$$

$$a^{1/y} = \sqrt[y]{a}$$

$$a^{x/y} = \sqrt[3]{a^x}$$

$$a^{-x} = \frac{1}{a^x} \text{ for } a \neq 0$$

$$a^0 = 1 \text{ for } a \neq 0$$

$$a^1 = a$$

$$\log_{10} x = \log x$$

$$\log_e x = \ln x$$

$$\log_a xy = \log_a x + \log_a y$$

$$\log_a \frac{x}{y} = \log_a x - \log_a y$$

$$\log_a x^m = m \log_a x$$

BINOMIAL THEOREM

$$(a \pm b)^n = a^n \pm na^{n-1}b + \frac{n(n-1)a^{n-2}b^2}{2!} \pm \frac{n(n-1)(n-2)a^{n-3}b^3}{3!} + $$

$$\cdots$$

QUADRATIC FORMULA

For $ax^2 + bx + c = 0$, $a \neq 0$, the roots of the equation are given by

$$x = \frac{-b \pm \sqrt{b^2 - 4ac}}{2a}.$$

LOGARITHMS AND EXPONENTIAL FUNCTIONS

Logarithms and exponential functions are widely used in engineering analysis. Very often there is some confusion as to what logarithms are, where they come from, and how they relate to exponential functions.

We start by representing a number, such as 123, in terms of another number, say 10, raised to a power:

$$10^{2.0899} = 123.$$

This is the basis of logarithms: denoting a number by a power of ten (in this case). Thus,

$$\log_{10} 123 = 2.0899$$

means that the log to the base 10 of 123 is 2.0899. In this text we adopt the conventional designation that log 123 implies a log to the base 10. This is sometimes called the common logarithm. Another frequently used logarithm is to the base e. This is represented as

$$\log_e x = \ln x.$$

In this text, $\ln x$ is used to denote a natural, or Napierian, logarithm, where $e = 2.71828 \ldots$. If we use the number 123 again, then

$$e^{4.81218} = 123,$$

or

$$\ln 123 = 4.81218.$$

The exponential character is more useful when our expression uses variables, rather than numbers; for example, $\ln x = y$, hence $x = e^y$. This curve is plotted in Figure A6.1. Notice that the curve is the same whether the natural logarithm or exponential form of the function is used.

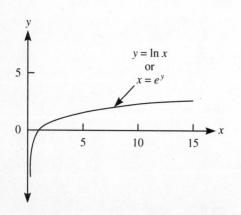

FIGURE A6.1 The plot of an exponential curve.

APPENDIX SEVEN

TRIGONOMETRY

This appendix contains some of the trigonometric functions that are commonly used in engineering calculations. It includes only functions from plane trigonometry. Figure A7.1 is a right triangle used to define various trigonometric functions.

$$\text{sine } \theta = \sin \theta = \frac{\text{opposite side}}{\text{hypotenuse}} = \frac{y}{r}$$

$$\text{cosine } \theta = \cos \theta = \frac{\text{adjacent side}}{\text{hypotenuse}} = \frac{x}{r}$$

$$\text{tangent } \theta = \tan \theta = \frac{\text{opposite side}}{\text{adjacent side}} = \frac{y}{x}$$

$$\text{cotangent } \theta = \cot \theta = \frac{\text{adjacent side}}{\text{opposite side}} = \frac{x}{y} = \frac{1}{\tan \theta}$$

$$\text{secant } \theta = \sec \theta = \frac{\text{hypotenuse}}{\text{adjacent side}} = \frac{r}{x} = \frac{1}{\cos \theta}$$

$$\text{cosecant } \theta = \csc \theta = \frac{\text{hypotenuse}}{\text{opposite side}} = \frac{r}{y} = \frac{1}{\sin \theta}$$

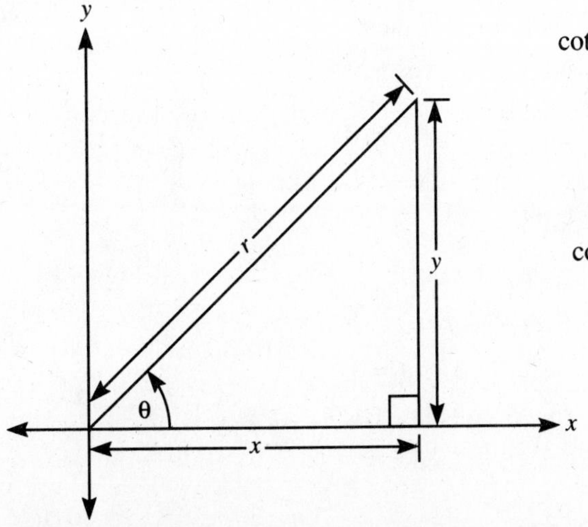

FIGURE A7.1 A right triangle.

TABLE A7.1 TRIGONOMETRIC SIGN VALUES IN THE FOUR QUADRANTS

Quadrant II		Quadrant I	
$x(-)$	$y(+)$	$x(+)$	$y(+)$
$\sin(+)$	$\csc(+)$	$\sin(+)$	$\csc(+)$
$\cos(-)$	$\sec(-)$	$\cos(+)$	$\sec(+)$
$\tan(-)$	$\cot(-)$	$\tan(+)$	$\cot(+)$
Quadrant III		**Quadrant IV**	
$x(-)$	$y(-)$	$x(+)$	$y(-)$
$\sin(-)$	$\csc(-)$	$\sin(-)$	$\csc(-)$
$\cos(-)$	$\sec(-)$	$\cos(+)$	$\sec(+)$
$\tan(+)$	$\cot(+)$	$\tan(-)$	$\cot(-)$

These trigonometric functions have positive and negative values depending on which quadrant they are located. Table A7.1 illustrates the function sign values.

In engineering calculations you must be able to represent angles in degrees or in radians. There are 2π radians in a circle or $360°$, hence

$$1° = \frac{2\pi}{360} = \frac{\pi}{180} = 0.017\ 453\ .\ .\ .\ \text{radians}$$

$$1\ \text{rad} = \frac{360}{2\pi} = \frac{180}{\pi} = 57.295\ 78\ .\ .\ .°.$$

Some trigonometric identities, listed below, are useful in reducing the complexity of equations.

$$\sin(-\theta) = -\sin\theta$$

$$\cos(-\theta) = \cos\theta$$

$$\sin\left(\frac{\pi}{2} \pm \theta\right) = \cos\theta$$

$$\cos\left(\frac{\pi}{2} \pm \theta\right) = \mp\sin\theta$$

$$\sin(\pi \pm \theta) = \mp\sin\theta$$

$$\cos(\pi \pm \theta) = -\cos\theta$$

$$\sin^2\theta + \cos^2\theta = 1$$

$$\sin 2\theta = 2\sin\theta\cos\theta$$

$$\cos 2\theta = 2\cos^2\theta - 1$$

$$\sin \frac{\theta}{2} = \sqrt{\frac{1 - \cos \theta}{2}}$$

$$\cos \frac{\theta}{2} = \sqrt{\frac{1 + \cos \theta}{2}}$$

$$\sin (\alpha \pm \beta) = \sin \alpha \cos \beta \pm \cos \alpha \sin \beta$$

$$\cos (\alpha \pm \beta) = \cos \alpha \cos \beta \mp \sin \alpha \sin \beta$$

Refer to Figure A7.2.

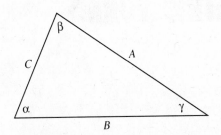

FIGURE A7.2 An acute triangle.

LAW OF SINES

$$\frac{A}{\sin \alpha} = \frac{B}{\sin \beta} = \frac{C}{\sin \gamma}$$

LAW OF COSINES

$$A^2 = B^2 + C^2 - 2BC \cos \alpha$$

CALCULATION OF LENGTH, AREA, AND VOLUME FOR VARIOUS GEOMETRIC FORMS

In this appendix, A = area, V = volume, and a, b, c, and h denote length.
For the oblique triangle illustrated in Figure A8.1,

$$A = \frac{1}{2}bh.$$

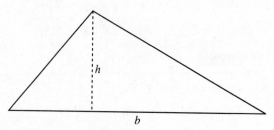

FIGURE A8.1 An oblique triangle.

For the parallelogram illustrated in Figure A8.2,

$$A = bh.$$

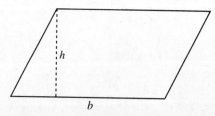

FIGURE A8.2 A parallelogram.

For the trapezoid illustrated in Figure A8.3,

$$A = \frac{1}{2} h(a + b).$$

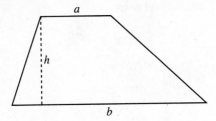

FIGURE A8.3 A trapezoid.

Refer to the circle illustrated in Figure A8.4, where the angle θ is measured in radians.

$$\theta = c/r$$

$$\cos \frac{\theta}{2} = \frac{b}{r}$$

$$a = r \sin \frac{\theta}{2}$$

$$A = \pi r^2 \qquad 2r = d \qquad A = \frac{\pi}{4} d^2$$

$$\text{Circumference} = 2\pi r = \pi d$$

$$\text{Area of sector } MNO = \frac{r^2\theta}{2}$$

$$\text{Area of triangle } MNO = br \sin \frac{\theta}{2}$$

$$\text{Area of shaded segment} = \frac{r^2\theta}{2} - br \sin \frac{\theta}{2}$$

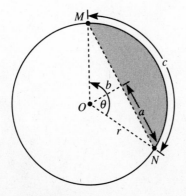

FIGURE A8.4 A circle.

For a sphere of radius r,

$$A = 4\pi r^2, \qquad V = \frac{4}{3}\pi r^3.$$

For a cylinder of radius r and height h,

$$V = \pi r^2 h,$$

$$A_{\text{side}} = 2\pi rh.$$

Refer to the cone in Figure A8.5.

$$V = \frac{\pi r^2 h}{3},$$

$$A_{\text{lateral}} = \pi rs$$

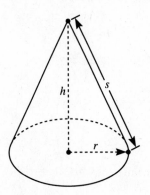

FIGURE A8.5 A cone.

PHYSICAL CONSTANTS

Avogadro's number: 6.022×10^{23} particles/mole

Density of dry air (15°C, 1 atm): 1.225 kg/m^3

Density of water (4°C): 1000 kg/m^3

Earth's average density: 5522 kg/m^3; 344.7 lbm/ft^3

Earth's equatorial radius: 6378 km; 3963 mi

Earth's mass: 5.976×10^{24} kg

Gravitational acceleration (standard at sea level): 9.806 m/s^2; 32.174 ft/sec^2

Gravitational constant: 6.672×10^{-11} N $\cdot$ m^2/kg^2

Heat of fusion of water (0°C): 333.8 kJ/kg; 143.5 Btu/lbm

Heat of vaporization of water (100°C): 2259 kJ/kg; 971.2 Btu/lbm

Molar gas constant: 8.314 kJ/kgmol-K

Planck's constant: 6.626×10^{-34} J/Hz

Velocity of light in a vacuum: 2.998×10^8 m/s

Velocity of sound in dry air (0°C, 1 atm): 331.5 m/s; 1088 ft/sec

DENSITIES AND SPECIFIC GRAVITIES OF SELECTED MATERIALS

Material	Specific Gravity	Average Density	
		kg/m^3	lbm/ft^3
Gases at 0°C *and* 1 atm			
Air		1.284	0.080 18
Carbon dioxide		1.977	0.1234
Carbon monoxide		1.251	0.078 06
Ethane		1.357	0.084 69
Helium		0.1784	0.011 14
Hydrogen		0.089 88	0.005 61
Methane		0.7176	0.0448
Nitric oxide		1.338	0.083 53
Nitrogen		1.251	0.078 07
Oxygen		1.429	0.089 21
Sulfur dioxide		2.927	0.1827
Liquids at 20°C			
Alcohol, ethyl	0.80	802	50.1
Alcohol, methyl	0.81	809	50.5
Benzene	0.88	880	54.9
Engine oil	0.89	888	55.4
Gasoline	0.67	670	41.8
Kerosene	0.80	801	50.0
Sea water	1.02	1025	64.0
Water	1.00	1000	62.4

Material	Specific Gravity	Average Density	
		kg/m^3	lbm/ft^3
Metals at 20°C			
Aluminum	2.64	2642.	164.9
Brass (cast)	8.53	8553	533.9
Bronze	8.17	8170	510.0
Copper (cast)	8.91	8910	556.2
Gold (cast)	19.30	19 300	1204.7
Iron (cast)	7.08	7080	441.9
Lead	11.37	11 370	709.7
MAnganese	7.61	7610	475.0
Mercury	13.59	13 590	848.3
Nickel	8.60	8600	536.8
Silver (cast)	10.50	10 500	655.4
Steel, cold drawn	7.83	7830	488.8
Tin (cast)	7.35	7350	458.8
Titanium	4.50	4500	280.9
Tungsten	19.20	19 200	1198.5
Uranium	18.70	18 700	1167.3
Zinc (cast)	7.05	7050	440.1
Nonmetals at 20°C			
Asbestos	2.45	2450	152.9
Brick (common)	1.80	1800	112.4
Cement, Portland	1.51	1505	93.9
Clay (damp)	1.80	1800	112.4
Coal (bituminous)	1.35	1350	84.3
Concrete	2.30	2300	143.6
Earth (loose)	1.20	1200	74.9
Earth (packed, moist)	1.50	1500	93.6
Glass (common)	2.60	2600	162.3
Gravel (loose)	1.60	1600	99.9
Gypsum	2.30	2300	143.6
Limestone	2.50	2500	156.1
Marble	2.75	2750	171.7
Masonry	2.40	2400	149.8
Paper	9.30	9300	580.5
Rubber	9.40	9400	586.8
Salt	2.15	2150	134.2
Sand (loose)	1.50	1500	93.6
Sugar	1.60	1600	99.9
Sulfur	2.00	2000	124.8
Wood (Douglas fir, dry)	0.51	510	31.8
Wood (maple, dry)	0.69	690	43.1
Wood (redwood, dry)	0.42	420	26.2
Wood (southern pine, dry)	0.64	640	40.0
Wood (white oak, dry)	0.77	770	48.1

CHAPTER SIX

6.13 $Y = 0.075 + 0.00575F$

6.19 $T = -10 + 3t$

6.21 $W = 2.511D^{1.975}$

6.29 $W = 90.1 + 6.553F$

6.31 $Y = 19.8\ X^{6.438}$

CHAPTER SEVEN

7.1 $R = 6; x_m = 9; \bar{x} = 9.71$

7.3 2 modes; $x_m = 75; \bar{x} = 73$

7.5 $\bar{x} = 186.8$ cars/h; $x_m = 160$ cars/h

7.7 5 modes; $\bar{x} = 12.5; x_m = 12.5$

7.9 $\bar{x} = 12.5$

7.11 $\sigma_1 = 1.98; \sigma_2 = 2.49$

7.13 $\sigma = 95.54$

7.15 $\sigma = 64.8$

7.17 <5 kg, 5000 bags; >5.1 kg, 1585 bags; >4.9 kg and <5.0 kg, 3415 bags

7.19 $\bar{x} = 19.17; \sigma = 0.57$

7.21 $T = 42.35x + 57.2°F$

7.23 1.6%

CHAPTER EIGHT

8.1 20 postcards, 30 letters

8.3 (a) 5.6192×10^2 (b) 8.31×10^0 (c) 3.96×10^{-2}
(d) 3.10125×10^3 (e) 7.29034×10^5 (f) 9.3×10^4
(g) 4×10^{-4} (h) 4.310×10^1

8.5 (a) 30.0 (b) 23.541 (c) 6.18×10^3 (d) 0.884 (e) 0.367

8.7 (a) 57.2°C (b) 9.78°F (c) 328.9°K (d) 5.5°K (e) −267.5°C
(f) 122°F (g) 581.67°R

8.9 $1920

8.11 14,400 lbm/min

8.13 9.87 m/s²; 32.38 ft/sec²

CHAPTER NINE

9.1 (a) 16.5 ohms (b) 6.67 ohms (c) 9.375 ohms

9.3 0.15 A

9.5 (a) 1.25 ohms (b) 20 ohms

9.7 24.46 ohms

9.9 16 bulbs

9.11 No current flow

9.13 151.8 ohms

9.15 1 mF

9.17 1.1 C

9.19 35.9 s

9.21 $R/L = 2$

9.25 22.5 sin 2500t mA

CHAPTER TEN

10.1 R = 72.7 N, 21.4°

10.3 **(a)** $R = 69.2$ N, 211.2° **(b)** $R = 39.5$ lbf, 190.3° **(c)** $R = 82.6$ N, $-19.7°$ **(d)** $R = 108.4$ N, 93°

10.5 $AC = 13,252$ N, $BC = 9,766$ N

10.7 0

10.9 7 N · m

10.11 1458.3 N↑, 1041.7 N↑

10.13 −25 N←, 43.25 N↑, 3.45 N↑

10.15 234.4 lbf↑, 20.6 lbf↑

10.17 $\sigma = 1273.2$ psi for all; $\epsilon_s = 4.39 \times 10^{-5}$, $\epsilon_a = 1.27 \times 10^{-4}$, $\epsilon_b = 9.79 \times 10^{-5}$

10.19 33,953 kPa, 169,765 kPa, $\Delta L = 0.34$ mm

10.21 10,000 kPa

10.23 4.47 cm

10.25 0.195 in.

10.27 12 mi/h

10.29 1000 N

10.31 8.63 m

10.33 120.4 m, 4.96 s

10.35 122.5 m, 24.5 s

10.37 9.14 cm

10.39 It moves

10.41 $R_1 = -375$ N↓, $R_2 = 375$ N↑, $R_3 = 1000$ N←

10.43 Same as 10.41

CHAPTER ELEVEN

11.1 25.5 lbm/sec, 183.6 gpm

11.3 500 kg/s, 254.6 m/s

11.5 19.24 kg/s, 110.9 m/s

11.7 31.9°C

11.9 5.7°C

11.11 (a) 2.4×10^7 Btu   (b) 1.342×10^5 Btu   (c) 2.873×10^5 kJ

11.13 77 kW

11.15 K.E. = 3.858×10^5 kJ; P.E. = 196 000 kJ

11.17 $m = 70$ kg

11.19 164.15 J; 3.6 m/s

11.21 -121.1 kW

11.23 70.3°F

11.25 46 305 kW

11.27 4.64×10^8 Btu/h

11.29 $\dot{Q}_{out} = 567$ MW; 214 715 gpm

11.31 5823 kg SO_2/day, 395 cars/week

11.33 2.51×10^8 kg

11.35 8.63×10^8 kg

11.37 74.3 years; 40 years

CHAPTER TWELVE

12.1 $1120

12.3 $12,000

12.5 $99,575

12.7 (a) $395/month   (b) $92.42/month

12.9 $143.47, $120.02, $110.11, $105.32, $102.86

12.11 $736.5/quarter

12.13 $P_o = -\$4237$, EAC = $-\$750$

12.15 AW = $-\$7351$, MARR = 8.26%

12.17 $i = 15.4\%$

12.19 2 levels

12.21 Case 2 least EAC; EAC = $26,939

12.23 $B/C = 0.829$

12.25 Case 3 least EAC; EAC = $3391

CHAPTER THIRTEEN

13.3 (a) $100101000_2 = 296_{10}$ (b) $100001000101101_2 = 16941_{10}$
(c) $11.0010010001_2 = 3.1416_{10}$ (d) $0.00011100011_2 = 0.1110_{10}$

13.5

Decimal	Octal	Hexadecimal
19	23	13
2015	3737	7DF
92.20	134.14	5C.3
0.824	0.645	0.D28

13.7 (a) 1110   (b) 1100   (c) 1111   (d) 100100000

13.9 (a) 110100   (b) 101100000   (c) 10010   (d) 11100011111